HEATING
AND HOT-WATER SERVICES

HEATING
AND
HOT-WATER SERVICES

Selected subjects with worked examples in S.I. Units

E. W. SHAW

F.I.H.V.E., F.R.S.H., F.I.PLANT E.
Senior Lecturer in Heating and Ventilating
National College for Heating, Ventilating,
Refrigeration and Fan Engineering

CROSBY LOCKWOOD & SON LTD
26 OLD BROMPTON ROAD · LONDON · SW7

© 3rd edition E. W. Shaw
First published 1963
Second edition 1966
Third edition 1970

ISBN 0 258 96801 X

*Typeset at the Pitman Press, Bath and printed in Great Britain by
Fletcher & Son Ltd, Norwich*

Preface to the First Edition

It is impossible in a book of moderate size to cover adequately all the subject matter that appears in recognized syllabuses on heating and hot-water services. In this book I have avoided purely descriptive topics and restricted the text to subjects that require an analytical treatment. The book therefore deals with fundamental principles and their application, and is intended to serve as a basic reference for practising engineers and as a text for students.

Although no particular examination syllabus has been followed, the book should be of service to students preparing for the examinations of The Institution of Heating and Ventilating Engineers, The National College for Heating, Ventilating, Refrigeration and Fan Engineering, The City and Guilds of London Institute and Schools of Architecture and Building and also University degrees in Building Engineering Services.

While the student must acquire a sound knowledge of the principles involved, he must also appreciate the limitations which arise in their application. If assumptions are made at the outset of a design they should subsequently be checked to show their validity. This can be done only if the principles involved have been previously exercised and understood. Each chapter therefore is devoted to a particular subject, and begins with an introduction in which relevant fundamental principles and rational procedures are developed or given. These are then illustrated by a number of fully worked-out examples, and each chapter ends with a selection of problems for the reader to solve. The symbols used are as far as possible in accordance with B.S. 1991, Part 1, 1954 and its amendments. All calculations have been made with the aid of a slide rule, and every effort has been made to avoid mistakes. If any errors have, however, gone undetected it is hoped that readers will call attention to them. I shall also be very grateful to receive from readers any suggestions for improvements in the text.

The properties of water in its liquid and vapour phases that have been used throughout the book are taken from the abridged set of steam tables included in *Thermodynamic Properties of Fluids and Other Data*, by Y. R. Mayhew and G. F. C. Rogers (Blackwell, Oxford). Problems marked (IHVE) and (NC) are taken with permission from recent examination papers of The Institution of Heating and Ventilating Engineers and The National College for Heating, Ventilating, Refrigeration and Fan Engineering respectively. It must be understood that these bodies are responsible only for the questions indicated and not for the answers. To follow the worked examples and attempt the problems students require a knowledge of Algebra up to quadratic equations, and should be able to differentiate x^n and integrate $x^n dx$.

I am indebted to The Institution of Heating and Ventilating Engineers for permission to quote and use some copyright material from the I.H.V.E. *Guide to Current Practice*, 1959 Edition, and to the British Standards Institute for permission to reproduce data from Code of Practice 342:1950 on Centralised Domestic Hot Water Services.

Experience gained during many years in technical teaching has undoubtedly influenced the writing of this book, and I am aware of the debt to my colleagues and former tutors for some of the ideas used.

Finally, I must express my thanks to my wife for her constant encouragement and material help during the preparation of the manuscript.

E. W. SHAW

Addington, May 1963

Preface to the Second Edition

The author takes this opportunity of thanking his colleagues and students for their interest in the first edition and the many other people previously unknown to him who have been good enough to write making suggestions for revisions or improvements.

A new chapter on elementary heat transfer has been included to help those readers who require only a general knowledge of this most important subject. It will be particularly useful to students taking the new City and Guilds of London Institute technician draughtsman's course in heating and ventilating engineering. The chapter on heat emission has been enlarged to include some general design data on embedded pipes and cables and radiant panels and typical design calculations.

The chapter on high-pressure hot water now includes a more detailed analysis of gas pressurization, further details of pressurization by head tank and some notes on the use of cascade type direct contact water heaters.

Considerably more data has been included on the use of pumps and steam traps.

It is regrettable that it has not been possible within the bounds of the present volume to include descriptions of plant and equipment requested by some readers of the first edition.

E. W. SHAW

Addington, March 1966

Preface to the Third Edition

Since publication of the earlier editions it has been agreed by the
Government (Hansard Vol 713, No. 121, Cols. 32–33) that there should
be a progressive changeover to metric units and that within ten years the
greater part of the country's industry will have effected the change. In
addition the International Organization for Standardization (ISO) has been
endeavouring to secure universal agreement for a single coherent system
and has now obtained a wide measure of agreement for the adoption of the
Système International d'Unités (SI). This system has already been accepted
as the legal system in many countries and in the U.K. is gradually replacing
the existing Imperial system. The British Standards Institution is already
preparing specifications incorporating SI metric units and it is expected
that within a year or two well over a thousand standards will have been
established in the new system. It has therefore been decided to adopt SI
metric units for this third edition, the aim being to provide the reader with
an insight into their application to practical problems in heating engineering.

While it follows from the Government's announcement that all syllabuses
and examinations will eventually need to be based on SI metric units the
adoption of SI will not affect the basic principles taught in heating engineering since these can mostly be presented without reference to any particular
system of units. However, experience gained during many years of teaching
has shown that worked examples are essential to illustrate the application
of basic principles to practical design problems. Although this edition is
essentially the second edition rewritten in SI metric units the opportunity
has been taken to revise the chapter on the heat requirements of heated
buildings and the chapter on hot water supply.

The adoption of SI metric units will involve changes not only in the
U.K. but also on the Continent and inevitably there will be a period of
transition when both the old and new systems will be in use. Readers are
therefore advised to refer to the British Standards Institution publication
The Use of SI Units (PD 5686) and the National Physical Laboratory's
booklet *Changing to the Metric System* published by HMSO. In addition
relevant conversion data are given in the IHVE publication *Change to
Metric*.

E. W. SHAW

Addington, February 1970

Contents

Principal Symbols Used

A	area
c	specific heat capacity; velocity of light
C	velocity factor; a constant
d	diameter
E	expansion volume; radiant emissive power
f	factor; function
F	radiation factor
g	acceleration due to gravity
G	mass velocity
h	surface conductance; heat transfer coefficient; height; specific enthalpy
I	intensity of radiation
k	velocity pressure factor
L, l	length; element of thickness
$\dot{m}$	mass flow rate
m	mass
n	index of compression or expansion; rotational speed
p	pressure
Q	quantity of heat
r	radius; thermal resistivity
R	thermal resistance; gas constant
t	common temperature
T	absolute temperature
u	velocity
U	overall coefficient of heat transfer
v	specific volume
V	volume
x	by-pass quantity; movement of water level; dryness fraction
Z	steam pressure factor
Δp	pressure difference
Δt	temperature difference

a absorptivity; thermal diffusivity
β coefficient of cubic expansion
ϵ emissivity
η efficiency; absolute viscosity
θ angle; time
λ wavelength; thermal conductivity
ν frequency
ρ density
σ Stefan-Boltzman constant
τ transmissivity
ϕ fin efficiency
Φ heat flow rate
Ω solid angle

Subscripts:

a air space; ambient air; absorbed
i inside
m mean
o outside
p pressure
r reflected; heat transfer by radiation
s surface; steam
c heat transfer by convection
1 case I
2 case II
f saturated liquid
g saturated vapour
fg change of phase at constant pressure
A configuration factor
E emissivity factor
e equivalent length

Introduction to Elementary Heat Transfer

Conduction

Consider Fig. (Int.1), which shows a thin parallel-sided single slab of homogeneous material having an area A and a thickness δl. Let side 1 be maintained at constant uniform temperature t and side 2 at a lower temperature $(t - \delta t)$, then in this steady state heat will flow at a steady rate from side 1 to side 2 in a direction normal, i.e. at right angles, to the faces of the slab. Under these conditions the rate of heat flow is found to be directly proportional to the area A and to the temperature fall $-\delta t$ and inversely proportional to the thickness δl, i.e.:

$$\text{Rate of heat flow } (\Phi) \propto A\delta t/\delta l$$

or $$\text{Rate of heat flow } (\Phi) = \lambda A\delta t/\delta l$$

where λ is a proportionality factor known as the thermal conductivity.

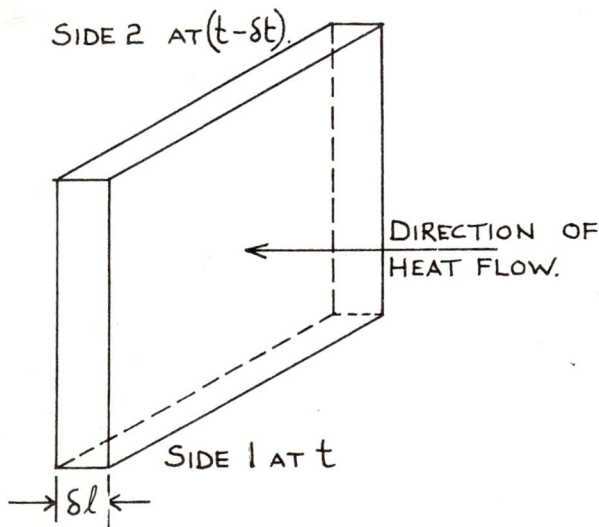

Fig. Int.1

1

The ratio $-\delta t/\delta l$ is the temperature gradient, and if δl is very small it can be written $-dt/dl$. If δQ is the quantity of heat passing in time $\delta\theta$ the rate of heat flow is $\delta Q/\delta\theta$ or $dQ/d\theta$ in the limit when $\delta\theta$ is very small. Hence for unidirectional conduction:

$$\frac{dQ}{d\theta} = -\lambda A \frac{dt}{dl} \tag{Int.1}$$

This simple relation is known as Fouriers law, and was first proposed in 1822. During the steady state the temperature varies with position only and not with time. It follows that the rate of heat flow $dQ/d\theta$ is also independent of time, hence $dQ/d\theta$ equals Q/θ and Eq (Int.1) may be written:

$$\Phi = -\lambda A \frac{dt}{dl} \tag{Int.2}$$

where Φ is the steady rate of heat flow.

From Eq (Int.2)

$$\Phi dx = -\lambda A \, dt$$

then integrating for a slab having a thickness l with faces at t_1 and t_2

$$\int_0^l \Phi dx = -\int_{t_1}^{t_2} \lambda A \, dt$$

Since A is constant

$$\Phi l = -A \int_{t_1}^{t_2} \lambda \, dt$$

If λ is assumed to be constant, then

$$\Phi l = -\lambda A \int_{t_1}^{t_2} dt$$

and

$$\Phi = -\frac{\lambda A}{l}\left(t_2 - t_1\right) \tag{Int.3}$$

or

$$\Phi = \frac{\lambda A}{l}\left(t_1 - t_2\right) \tag{Int.4}$$

from which

$$\lambda = \Phi l / A(t_1 - t_2) \tag{Int.5}$$

If Φ is expressed in W; A in m^2; l in m and t in °C the units of λ will be:

$$\frac{Wm}{m^2{}^\circ C}, \text{ i.e. } \frac{W}{m{}^\circ C}$$

It should be noted that the thermal conductivity of any particular solid material is rarely a constant but a function of its mean temperature, density and moisture content, and generally increases more or less linearly with these properties. With liquids and gases the thermal conductivity depends also upon the pressure. The thermal conductivity may therefore be considered to be a property of the material through which the heat flows, and although basically defined by Eq (Int.5), is evaluated experimentally. Methods of determining thermal conductivity are described in B.S. 874, which should be referred to for specific information. The thermal conductivity of a wide range of materials used in building construction and heating and ventilating engineering is given in *A Guide to Current Practice 1970*, published by the Institution of Heating and Ventilating Engineers, from which the data given in Table 1.1 have been taken. A material which has a low thermal conductivity offers a high resistance to the flow of heat and may be considered to be a thermal insulator. The reciprocal of the thermal conductivity gives the specific resistance, or resistivity (r), of the material. See Table Int.1.

Conduction through Composite Walls

Consider steady heat flow through the composite wall shown in Fig. Int.2, which is made up of two parallel-sided slabs 1 and 2 having thicknesses l_1 and l_2 and thermal conductivities λ_1 and λ_2 respectively. Let the temperature of the outer faces of the slabs be t_1 and t_3 and let t_2 be the temperature of the interface. Assume that t_1 is greater than t_3 and that heat flows normal to the faces, i.e. let the face area A be large compared to l, then from Eq (Int.4):

$$\Phi = \frac{\lambda_1 A}{l_1}(t_1 - t_2) = \frac{\lambda_2 A}{l_2}(t_2 - t_3)$$

from which

$$t_2 = \frac{\dfrac{\lambda_1}{l_1} t_1 + \dfrac{\lambda_2}{l_2} t_3}{\dfrac{\lambda_1}{l_1} + \dfrac{\lambda_2}{l_2}}$$

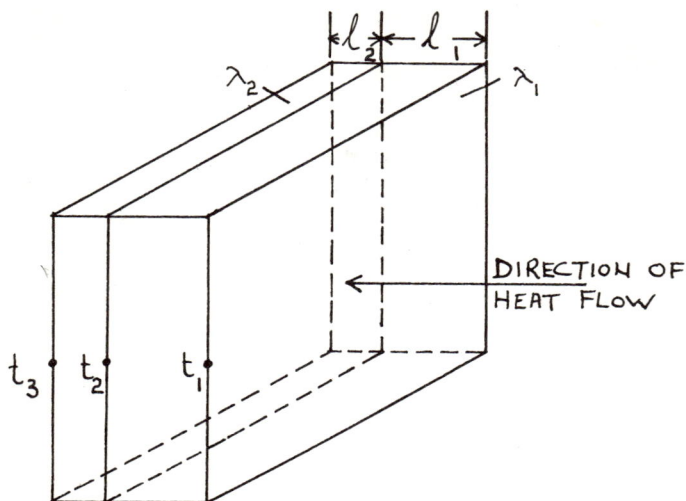

Fig. Int.2.

Since

$$t_1 - t_2 = \Phi \frac{l_1}{\lambda_1 A}$$

and

$$t_2 - t_3 = \Phi \frac{l_2}{\lambda_2 A}$$

then adding these two equations,

$$t_1 - t_3 = \Phi \left(\frac{l_1}{\lambda_1 A} + \frac{l_2}{\lambda_2 A} \right)$$

$$\therefore \ \Phi = \frac{t_1 - t_3}{\dfrac{l_1}{\lambda_1 A} + \dfrac{l_2}{\lambda_2 A}}$$

or

$$\frac{\Phi}{A} = \frac{t_1 - t_3}{\dfrac{l_1}{\lambda_1} + \dfrac{l_2}{\lambda_2}} \qquad\qquad (Int.6)$$

Alternatively, if r_1 and r_2 represent the thermal resistivities of the slabs:

$$\frac{\Phi}{A} = \frac{t_1 - t_3}{r_1 l_1 + r_2 l_2} \qquad\qquad (Int.7)$$

The terms $r_1 l_1$ and $r_2 l_2$ in Eq (Int.7), and $\dfrac{l_1}{\lambda_1}$ and $\dfrac{l_2}{\lambda_2}$ in Eq (Int.6) are called thermal resistances (R). It is clear from these equations that the thermal

TABLE Int.1

Thermal Conductivity and Resistivity of Building Materials

Material	Moisture content, (% of dry weight)	Density, kg/m^3	Conductivity λ $W/m°C$	Resistivity $r = 1/\lambda$, $m°C/W$
Asbestos insulating board . .	2	720	0.12	8.33
	2	1 200	0.25	4.00
Asphalt roofing	–	1 600	0.43	2.32
Brickwork:				
Common brickwall	3	–	1.15	0.87
London stock	–	–	0.79	1.26
Coke breeze slab	–	–	0.58	1.72
Concrete:				
Ballast: 1 : 2 : 4	–	2 241 – 2 480	1.44	0.69
Cellular	–	640	0.14	7.14
Fibreboard, insulating				
board	10–12	320	0.58	1.72
Foamed slag normal				
aggregate	4.7	1 088	0.25	4.00
Glass:				
Sheet, window	–	2 510	1.05	0.95
Hollow glass block wall .	–	–	0.65–0.72	1.54–1.39
Gypsum plaster	–	1 120	0.37	2.7
Gypsum plasterboard	–	960	0.16	6.25
Roofing felt	–	960	0.19	5.26
Stone:				
Granite	–	2 640	2.93	0.35
Limestone	–	2 180	1.53	0.65
Sandstone	–	2 000	1.3	0.77
Tiles:				
Burnt clay	–	1 920	0.84	1.19
Cork	–	528	0.08	12.50
Plastic	–	1 040	0.51	1.96
Timber:				
Deal	12	610	0.13	7.70
Oak	14	770	0.16	6.25
Pitch pine	15	656	0.14	7.14
Plywood	12	528	0.14	7.14
Wood chip board	–	800	0.14	7.14
Wood-wool cement slabs . .	5	640	0.12	8.35

resistance (R) of a material is directly proportional to its thickness. Eqs (Int.6) and (Int.7) may be written:

$$\frac{\Phi}{A} = \frac{t_1 - t_3}{R_1 + R_2} \qquad (Int.8)$$

The above method may be used to obtain the rate of heat flow and the interface temperatures when there are more than two slabs, and in general:

$$\frac{\Phi}{A} = \frac{\Delta t}{\Sigma R} \qquad (Int.9)$$

where

Φ = rate of heat flow, W

A = area, m^2

Δt = difference in temperature between the outer surfaces of the slab, °C

R = thermal resistance,

$$= r.l \text{ or} \frac{l}{\lambda}, \text{m}^2\text{°C/W}$$

and

r = thermal resistivity, m°C/W

λ = thermal conductivity, W/m°C

l = thickness, m

Note that Eq (Int.9) is similar to Ohm's law.

Convection

Convection takes place only in fluids, liquid or gases, and refers to the transfer of heat from one part of a fluid to another part at a lower temperature by actual movement of fluid particles. Practical heat-exchanger problems, however, are concerned mainly with the transfer of heat between solid surfaces and fluid heating media. These problems involve conduction as well as convection and take into account the difference in temperature between the solid surface and the main body of the fluid, the physical properties of the fluid and the nature of its movement over the surface. Experimental data show that the heat transfer by convection is related to the type of fluid flow, i.e. streamline and turbulent. There are two types of convection, namely: natural or free convection and forced convection. Natural convection takes place when the fluid flow is due only to the temperature, and hence density difference within the fluid, and therefore always takes place vertically. Forced convection occurs when the fluid is forced over the solid surface by mechanical means. With both types it is convenient to assume that a stationary film of fluid exists at the solid surface and that heat is conducted through this film either into or out of the main body of the fluid. The thickness of this film, and hence its resistance to conductive heat flow, is dependent on the fluid velocity. In the case of turbulent flow the film has negligible thickness and the heat transfer takes place mostly by convection. The rate of heat flow by convection between a surface and a fluid is given by:

$$\Phi_c = h_c A(t_s - t_m) \qquad (\text{Int.}10)$$

where

Φ_c = rate of heat flow by convection W

h_c = coefficient of heat transfer by convection W/m^2°C

t_s = surface temperature °C

t_m = mean temperature of fluid °C
A = heat transfer area m^2

Forced Convection. Since the precise fluid flow pattern and boundary conditions are not normally known, the coefficient of heat transfer by convection h_c in Eq (Int.10) is usually determined experimentally. In most cases the main variables affecting heat transfer are known and may be investigated by analysis of their dimensions; the heating effect due to fluid friction in simple hot water and steam systems is very small and is neglected in the analysis. The following variables are generally assumed to affect the coefficient of heat transfer h_c in forced convection:

η fluid dynamic viscosity
ρ fluid density
λ thermal conductivity of the fluid
c specific heat capacity of the fluid
Δt temperature difference between fluid and surface
u fluid velocity
l characteristic linear dimension

i.e. $h_c = f(\eta, \rho, \lambda, c, \Delta t, u, l)$ (Int.11)

where f is some function.
    Eq (Int.11) may be expanded into an infinite series of terms each of the form:

$$h_c = C\eta^a \rho^b \lambda^c c^d \Delta t^e u^f l^g \qquad (Int.12)$$

where C = a dimensionless constant and a, b, c, d, e, f and g are arbitrary indices.
    Since both sides of Eq (Int.12) must have the same dimensions we have:

$$\frac{Q}{L^2Tt} = \left(\frac{M}{LT}\right)^a \times \left(\frac{M}{L^3}\right)^b \times \left(\frac{Q}{LTt}\right)^c \times \left(\frac{Q}{Mt}\right)^d \times (t)^e \times \left(\frac{L}{T}\right)^f \times (L)^g$$

$$(Int.13)$$

i.e. collecting the powers of each fundamental dimension:

$$\frac{Q}{L^2Tt} = M^{(a+b-d)} \times L^{(-a-3b-c+f+g)} \times T^{(-a-c-f)} \times t^{(-c-d+e)} \times Q^{(c+d)}$$

The power to which each fundamental dimension is raised must be the same on both sides of Eq (Int.13), therefore by equating indices we have:

Q: $1 = c + d$
L: $-2 = -a - 3b - c + f + g$
T: $-1 = -a - c - f$
t: $-1 = -c - d + e$
M: $0 = a + b - d$

There are five equations and seven unknowns and solving in terms of d and f it will be found that:

$$a = (d - f); \quad b = f; \quad c = (1 - d); \quad e = 0 \text{ and } g = (f - 1)$$

Substituting for the indices in Eq (Int.12), a typical term in the series will be in the form:

$$h_c = C \eta^{(d-f)} \rho^f \lambda^{(1-d)} c^d \Delta t^o u^f l^{(f-1)}$$

i.e.

$$h_c = C \frac{\lambda}{l} \left(\frac{\eta c}{\lambda} \right)^d \left(\frac{\rho u l}{\eta} \right)^f$$

or since $\dfrac{\lambda}{l}$ is a common factor, in the form:

$$\frac{h_c l}{\lambda} = C \left(\frac{\eta c}{\lambda} \right)^d \left(\frac{\rho u l}{\eta} \right)^f \tag{Int.14}$$

which includes three dimensionless groups named after the original investigators as follows:

$$\frac{h_c l}{\lambda} = \text{Nusselt number} \qquad \text{(Nu)}$$

$$\frac{\rho u l}{\eta} = \text{Reynolds number} \qquad \text{(Re)}$$

$$\frac{\eta c}{\lambda} = \text{Prandtl number} \qquad \text{(Pr)}$$

i.e.

$$\text{Nu} = C \text{Pr}^d \text{Re}^f \tag{Int.15}$$

The value of the constant C and the two unknown powers d and f are found by experiment. For turbulent fluid flow (Re > 2 100) in long tubes with only small differences in temperature between the bulk fluid and the surface of the tube Eq (Int.14) becomes:

$$\frac{h_c d}{\lambda} = 0.023 \left(\frac{\rho u d}{\eta} \right)^{0.8} \left(\frac{\eta c}{\lambda} \right)^{0.33} \tag{Int.16}$$

in which d = diameter of the tube. It is frequently convenient to express Re in terms of the mass velocity $G = \dfrac{\dot{m}}{A}$, where $\dot{m}$ = mass flow = $uA\rho$, and A = area, i.e.

$$\frac{h_c d}{\lambda} = 0.023 \left(\frac{dG}{\eta} \right)^{0.8} \left(\frac{\eta c}{\lambda} \right)^{0.33} \tag{Int.17}$$

The fluid properties in Eq (Int.16) and Eq (Int.17) should be taken at the mean bulk temperature of the fluid. In cases where the difference in temperature between the surface and the bulk of the fluid is large, and for greater accuracy, the fluid properties should be taken at the arithmetic

TABLE Int.2

Properties of Air and Water

Property	Water				Air		
t °C	40	80	160	200	30	80	125
ρ kg/m³	994.6	974.1	909.7	866.7	1.177	0.998	0.883
η Ns/m²	6.5×10^{-4}	3.57×10^{-4}	1.72×10^{-4}	1.36×10^{-4}	1.85×10^{-5}	2.08×10^{-5}	2.39×10^{-5}
c J/kg°C	4 178	4 196	4 342	4 505	1 006	1 009	1 014
λ W/m°C	0.628	0.668	0.680	0.665	0.0262	0.0300	0.0337
Pr $\dfrac{\eta c}{\lambda}$	4.34	2.22	1.10	0.92	0.708	0.697	0.689

mean of the mean bulk temperature and the surface temperature, i.e. at the mean film temperature. Some properties of air and water are given in Table Int.2.

Natural Convection. The environmental engineer is frequently concerned with the natural convection that takes place at the surface of steam and hot-water pipes, air ducts, heating panels and room surfaces. Consider the case of a warm surface cooling in air. Heat flows by conduction from the surface into the adjacent layer of air, the rate of heat flow depending on the difference in temperature (Δt) and upon the thermal conductivity (λ) of the air. The corresponding rise in temperature produced in this layer will depend on the specific heat (c) and the resulting change in density (ρ) will depend on the volume coefficient (β) of the air. The density of the air adjacent to the warm surface is thus less than that of the main body of the air, and as a result, buoyant forces will cause an upward flow of air over the surface. The air will accelerate until the viscous resistance to its movement equals the net upthrust; the viscosity (η) is thus involved. In this way the heat conducted into the air adjacent to the surface is carried away in the natural convection currents. The rate of heat transfer will depend also on the shape and size of the warm surface and its position in space, i.e. horizontal or vertical. The heat transfer by radiation from the warm surface will be dealt with later.

If it is assumed that the rate of heat transfer at the surface due to natural convection alone is dependent on the following variables:

$$\lambda, l, \Delta t, c, \rho, \eta, \beta, g$$

it may be shown by dimensional analysis that:

$$Nu = C.f\{(Gr), (Pr)\} \qquad \text{(Int.18)}$$

where

    C = a constant
    f = some function
    Gr = Grashof number

$$Gr = \frac{\beta g \rho^2 l^3 \Delta t}{\eta^2}$$

where

    β = coefficient of cubical expansion
       = T^{-1} for ideal gases
    g = acceleration due to gravity
    ρ = density
    l = characteristic dimension
    Δt = difference in temperature between surface and fluid
    η = dynamic viscosity
    T = absolute temperature

The nature of the function (*f*) and the value of the constant (C) in Eq (Int.18) are determined experimentally. For the case of a single horizontal pipe freely exposed in draught-free air the heat-transfer coefficient for natural convection (h_c) has been shown* to be:

$$h_c = 0.53 \frac{\lambda}{D_o} (Gr.Pr)^{0.25} \qquad (Int.19)$$

where

h_c = heat transfer coefficient W/m²°C
λ = thermal conductivity W/m°C
D_o = outside diameter of pipe m
Gr = Grashof number, dimensionless
Pr = Prandtl number, dimensionless

The following approximate equations may be used with sufficient accuracy for practical cases when $10^4 < Gr < 10^9$.

Horizontal pipes:

$$h_c = 1.32 \left(\frac{\Delta t}{D_o}\right)^{0.25} \text{ W/m}^2°\text{C}$$

or

$$h_c = 7.42 \left(\frac{\Delta t}{d_o}\right)^{0.25} \text{ W/m}^2 °\text{C} \qquad (Int.20)$$

where

Δt = difference in temperature between surface and air, °C
D_o = outside diameter of pipe, m
d_o = outside diameter of pipe, mm

Plane surfaces:

$$h_c = C(\Delta t)^{0.25} \text{ W/m}^2°\text{C} \qquad (Int.21)$$

where

C = 1.3 W/m²°C¹·²⁵ for horizontal surfaces, heat flow down
1.9 ”   ”   ”   for vertical surfaces
2.5 ”   ”   ”   for horizontal surfaces, heat flow up

From Eq (Int.20) and Eq (Int.21) we have:

For horizontal pipes:

$$\frac{\Phi_c}{A} = \frac{7.42 \, \Delta t^{1.25}}{d_o^{0.25}} \text{ W/m}^2 \qquad (Int.22)$$

and for plane surfaces:

$$\frac{\Phi_c}{A} = C.\Delta t^{1.25} \text{ W/m}^2 \qquad (Int.23)$$

* McADAMS, W. H.: *Heat Transmission*, 3rd Edition, 1954, p. 177 (McGraw-Hill).

Mean Temperature Difference

The most common type of heat exchanger used in heating engineering is the recuperator, in which heat is transferred from one fluid to another through a solid wall such as a pipe wall. One of the fluids may be at constant temperature, as in the case of a condensing vapour or evaporating liquids, or the temperature of both fluids may vary as they flow over the heat-transfer surface. In all cases it is necessary to determine the true mean temperature difference between the two fluids in terms of their initial and final temperatures and relative flow paths, i.e. parallel, counter or cross flow. It may be shown that the true or logarithmic mean temperature difference (Δt_m) between two fluids in parallel or counterflow is:

$$\Delta t_m = \frac{\Delta t_{max} - \Delta t_{min}}{\log_e \frac{\Delta t_{max}}{\Delta t_{min}}} \qquad \text{(Int.24)}$$

where Δt = temperature difference between the two fluids, deg.C.

N.B. $\log_e N = 2.3026 \log_{10} N$.

Then

$$\Phi = U \cdot A \cdot \Delta t_m \qquad \text{(Int.25)}$$

where

U = overall coefficient of heat transfer $W/m^2\,^{\circ}C$
A = heat transfer area m^2

It should be noted that if the temperature of one of the fluids remains constant, then the change in temperature of the other will be the same whether the flow pattern is parallel or counterflow. If with counterflow arrangements, $\Delta t_{max} = \Delta t_{min}$, then Eq (Int.24) is indeterminate and it may be shown that $\Delta t_m = \Delta t_{max} = \Delta t_{min}$ and the arithmetic mean temperature difference may be used.

For a given heat load and heat-transfer coefficient the logarithmic mean temperature difference is greater for counterflow than for parallel-flow, and the heat-transfer area is therefore less, except when one fluid remains at constant temperature, when the logarithmic mean temperature difference is the same for both arrangements.

The rate of heat flow (Φ) is usually the only quantity that is known at the outset of a design. The designer must therefore determine U and Δt_m before the heat transfer area A can be determined. While the thermal transmittance U increases with increase in fluid velocity, so also does the pressure loss, and hence pumping power. The mass flow rates and operating temperatures must therefore be carefully selected to give optimum conditions and ensure minimum owning and operating costs.

Thermal Radiation

All matter continuously emits electromagnetic radiation in the form of rays covering a wide range of wavelengths, the quantity and spectral composition depending on the absolute temperature. For our purpose *thermal radiation* may be defined as the transfer of heat from one body to another at a lower temperature by electromagnetic waves passing through the intervening space which may be an absolute vacuum. Thermal radiation is similar in nature to other types of electromagnetic radiation such as light, radio waves and X-rays, differing only in wavelength (λ) and frequency (ν), and hence characteristic properties. All electromagnetic radiations travel in free space with the speed of light (c) ($2.997\,925 \times 10^8$ m/s), therefore $\nu = c/\lambda$. The wavelength of visible light ranges from 0.4 μm at the violet end of the

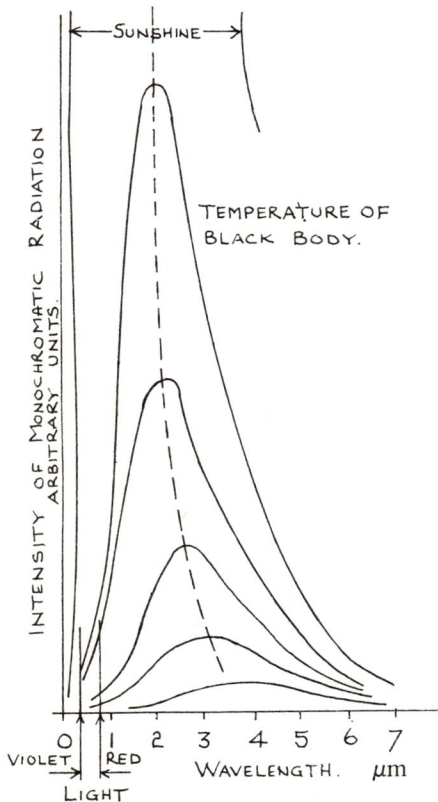

Fig. Int.3. Distribution of radiation from a black body
at different temperatures

spectrum to 0.8 μm at the red end. Analysis of a complete spectrum of sunlight shows that heat is being transferred as well as light. While a significant amount of thermal radiation occurs in the visible range of wavelengths, practical thermal radiation problems are concerned mostly with the near and far infra-red region.

For a given temperature there is a theoretical maximum amount of energy which can be emitted. Such a perfect radiator is commonly called a *black body*. The total radiation emitted by a black body is the sum of the radiation that takes place at all wavelengths of the body's spectrum. The distribution of energy for a black body at several temperatures is shown in Fig. (Int.3). It should be noted that at each temperature there is a wavelength at which the amount of energy given off is a maximum and that the wavelength at which the maximum occurs increases as the temperature of the body is reduced. It is also apparent that as the temperature increases the energy at each wavelength increases also. The radiation emitted at any one wavelength is called the monochromatic radiation, and the area under the curve represents the total radiation, i.e. the sum of the monochromatic values for the particular temperature. It is the total radiation that is usually required in the solution of practical problems.

Like light, thermal radiations travel in straight lines in a homogeneous medium; they have the same velocity as light and obey the same laws of reflection and refraction, and may also be polarized by concave mirrors and lenses, as in solar boilers. In general, thermal radiations are absorbed by dark rough surfaces and reflected by light smooth surfaces. When a ray of thermal radiation strikes a surface it may be partly reflected, absorbed or transmitted, depending on the nature of the surface. Let I = incident radiation and subscripts r, a and t refer to the reflected, absorbed and transmitted radiation respectively, then:

$$\text{Reflectivity } (\rho) = \frac{I_r}{I}$$

$$\text{Absorptivity } (a) = \frac{I_a}{I}$$

$$\text{Transmissivity } (\tau) = \frac{I_t}{I}$$

Since $$I = I_r + I_a + I_t$$

Then, for diathermanous materials,

$$\rho + a + \tau = 1 \qquad \text{(Int.26)}$$

Most solids and liquids are heat opaque, and Eq (Int.26) becomes

$$\rho + a = 1 \qquad \text{(Int.27)}$$

The amount of radiation absorbed must be dissipated if the temperature of the absorbing body is to remain constant. It is either lost from the surface by convection, re-radiated or conducted away, depending upon the ambient air temperature and the temperature and nature of the surrounding surfaces.

The amount of heat radiated from a surface per unit area per unit time is called the *emissive power* and denoted by the symbol E. If E_b is the emissive power of a black surface at the same temperature, then the ratio E/E_b is the emissivity (ϵ) of the surface. It can be shown (Kirchoff's law) that for most surfaces the absorptivity (a) of a body is equal to its emissivity (ϵ), provided that the wavelength of the incident radiation is similar to the wavelength of the emitted radiation, and also that the emissive power of an actual body must be less than that of a black body at the same temperature. An ideal body having a constant monochromatic emissivity at all wavelengths is called a *grey body*. The monochromatic emissivity of most actual surfaces is not constant at all wavelengths. The variation, however, is fairly small, and for many actual materials the emissivity is assumed constant for all wavelengths and temperatures. Typical emissivity factors for surfaces at about 24°C are given in Table (Int.3).

TABLE Int.3

Emissivity of Various Surfaces

(Surface Temperature about 24°C)

Surface	*Emissivity (ϵ)*
Asbestos board	0.96
Brickwork	0.9
Concrete	0.9
Glass, window	0.93
Metals:	
Polished brass, copper	0.04
Highly polished aluminium, tin plate, chromium	0.03
Dull aluminium, brass, copper, galvanized steel, polished iron	0.25
Rusty iron and steel	0.9
Paints:	
Non-metallic, all colours	0.9
Metallic, aluminium, bronze	0.5
Plaster	0.9
Stone	0.9
Tiles	0.9
Wood	0.9
Water	0.95

The heat emission by thermal radiation from a black body is given by the Stefan-Boltzman law:

$$\Phi_r/A = \sigma T^4 \qquad \qquad \text{(Int.28)}$$

where

Φ_r = heat emission by radiation, W
A = surface area, m^2
σ = Stefan-Boltzman radiation constant, i.e. 5.67×10^{-8} $Wm^{-2}K^{-4}$
T = absolute temperature, K

For an actual body the heat emission will be less, and if the body is assumed to be grey, then

$$\Phi_r/A = \sigma . \epsilon . T^4 \qquad \text{(Int.29)}$$

The net amount of radiation exchanged between two bodies depends not only on their relative emissivities and absorptivities but also on their shapes and relative orientations, and for most practical cases may be determined by applying the Stefan-Boltzman law as follows:

$$\Phi_r/A = \sigma . F_A . F_E (T_1^4 - T_2^4) \qquad \text{(Int.30)}$$

where

F_A = a dimensionless factor $\leqslant 1$, which takes into account the relative geometry of the two surfaces at T_1 and T_2
F_E = a dimensionless factor $\leqslant 1$, which takes into account the emissivities and absorptivities of the two surfaces at T_1 and T_2
A = surface area, see Table Int.4.

Eq (Int.30) is frequently written:

$$\Phi_r/A = 5.67 . F_A . F_E \left[\left(\frac{T_1}{100} \right)^4 - \left(\frac{T_2}{100} \right)^4 \right] \qquad \text{(Int.31)}$$

It may be assumed with sufficient accuracy for most practical cases that $F_A = 1$ for large parallel planes, long concentric cylinders and small bodies in large enclosures. The corresponding values of F_E are given in Table Int.4. The determination of F_A and F_E for particular cases is outside the scope of this book, and for detailed data reference should be made to specialized textbooks on heat transfer, such as *Heat Transmission*, by McAdams, W. H. (McGraw-Hill, 1954).

It is sometimes convenient to express the rate of heat transfer by thermal radiation as follows:

$$\Phi_r/A = h_r (t_1 - t_2) \qquad \text{(Int.32)}$$

where

h_r = radiation coefficient, $W/m^2 {}^\circ C$
t_1 and t_2 = surface temperatures, $^\circ C$

TABLE Int. 4

Typical Emissivity Factor and Related Areas for use in Eq (Int.31)

Case	Area	F_E
1. Large parallel planes	A_1 or A_2	$\dfrac{1}{\dfrac{1}{\epsilon_1} + \dfrac{1}{\epsilon_2} - 1}$
2. Long concentric cylinders	A_1^*	$\dfrac{1}{\dfrac{1}{\epsilon_1} + \dfrac{A_1}{A_2}\left(\dfrac{1}{\epsilon_2} - 1\right)}$
3. Small bodies in large enclosures	A_1^*	ϵ_1

From Eq (Int.31) and Eq (Int.32):

$$h_r = 5.67\, F_A\, F_E\left[\left(\frac{T_1}{100}\right)^4 - \left(\frac{T_2}{100}\right)^4\right](t_1 - t_2)^{-1} \qquad \text{(Int.33)}$$

or $\qquad h_r = 5.67 \times 10^{-8}\, F_A\, F_E(T_1^2 + T_2^2)(T_1 + T_2)$ $\qquad$ (Int.34)

For normal rooms a combined value of $F_A . F_E = 0.87$ may be used, and Eq Int.36 becomes

$$\Phi_r/A = 4.93\left[\left(\frac{T_1}{100}\right)^4 - \left(\frac{T_2}{100}\right)^4\right] \qquad \text{(Int.35)}$$

this is accurate to within 10 per cent and suitable for most practical cases.

Example Int.1. Calculate the rate of heat transfer by conduction across a furnace wall which has a thickness of 0.5 m and surface temperatures of 260°C and 27°C. The thermal conductivity of the wall is 1.08 W/m°C. From Eq (Int.4)

$$\Phi/A = \frac{1.08}{0.5}(260 - 27)$$

$$= 503 \text{ W/m}^2$$

Example Int.2. A heat exchanger is required to heat 0.64 kg/s of water from 60°C to 82°C. Calculate for (*a*) counterflow and (*b*) parallel-flow the heating surface required if 0.13 kg/s of high pressure hot water at 204°C is used as the primary heating medium. Assume that the overall heat-transfer coefficient is 1.14 kW/m²°C and take the specific heat capacity of the low and high temperature hot water as 4.19 and 4.27 kJ/kg respectively.

Let t_o = outlet temperature of the high-pressure hot water.

* Subscript 1 refers to the enclosed body.

Then, since the heat gained by the secondary medium must be equal to the heat given up by the primary medium, we have

$$\Phi = 0.64 \times 4.19(82 - 60) = 0.13 \times 4.27(204 - t_o)$$

from which, $t_o = 97.7°C$

and $\Phi = 0.64 \times 4.19(82 - 60) = 59$ kW

(*a*) For a counterflow arrangement, the temperature changes will be as shown below:

Primary medium $204°C \longrightarrow 97.7°C$
Secondary medium $82°C \longleftarrow 60.0°C$

$$\Delta t_{max} = 122°C \quad \Delta t_{min} = 37.7°C$$

From Eq (Int.24)

$$\Delta t_m = \frac{122 - 37.7}{\log_e \frac{122}{37.7}} = \frac{84.3}{1.1756} = 71.7°C$$

From Eq (Int.25), the heating surface required will be:

$$A = \frac{59}{1.14 \times 71.7} = 0.72 \text{ m}^2$$

(*b*) For a parallel-flow arrangement we have:

Primary medium $204°C \longrightarrow 97.7°C$
Secondary medium $60°C \longrightarrow 82.0°C$

$$\Delta t_{max} = 144°C \quad \Delta t_{min} = 15.7°C$$

From Eq (Int.24)

$$\Delta t_m = \frac{144 - 15.7}{\log_e \frac{144}{15.7}} = \frac{128.3}{2.2162} = 57.9°C$$

From Eq (Int.25), the heating surface required will be:

$$A = \frac{59}{1.14 \times 57.9} = 0.89 \text{ m}^2$$

i.e. $\frac{(0.89 - 0.72)100}{0.72} = 23.6$ per cent more than for the counterflow arrangement.

Example Int.3. Calculate the coefficient of heat transfer at the inside surface of a 4 cm internal diameter boiler tube through which water is flowing at a velocity of 0.6 m/s and a mean film temperature of 160°C.

The properties of water at 160°C are taken from Table Int.2, they are:

$$\lambda = 0.68 \text{ W/m}°\text{C}$$
$$\rho = 909.7 \text{ kg/m}^3$$
$$\eta = 1.72 \times 10^{-4} \text{ Ns/m}^2$$
$$\text{Pr} = 1.1$$
$$\text{Re} = \frac{\rho u d}{\eta}$$
$$= \frac{909.7 \times 0.6 \times 4 \times 10^{-2}}{1.72 \times 10^{-4}} = 127\,000$$

Since the flow is turbulent (Re > 2 100) Eq (Int.16) may therefore be used, i.e.

$$h_c = \frac{0.023 \times 0.68 \times 127\,000^{0.8} \times 1.1^{0.33}}{4 \times 10^{-2}}$$
$$= 4\,884 \text{ W/m}^2°\text{C}$$

Example Int.4. A pipe having an inside diameter of 8 cm is maintained uniformly at 100°C by an electric heating tape. What length of pipe is needed to warm water flowing through it at the rate of 0.4 kg/s from 50°C to 70°C.

Since there is a fairly large difference between the fluid and surface temperature, the properties of the water should be taken at the mean film temperature (t_f):

$$t_f = \tfrac{1}{2}(t_b + t_s)$$

where t_b = mean bulk temperature,

and t_s = surface temperature.

In this example,

$$t_f = \frac{1}{2}\left(\frac{50 + 70}{2} + 100 \right)$$
$$= 80°\text{C}$$

and from Table Int.2 the relevant properties are:

$$c = 4\,196 \text{ J/kg}°\text{C}$$
$$\eta = 3.57 \times 10^{-4} \text{ Ns/m}^2$$
$$\lambda = 0.668 \text{ W/m}°\text{C}$$
$$\text{Pr} = 2.22$$
$$\text{Re} = \frac{dG}{\eta} \text{ where } G = \frac{\dot{m}}{A}$$
$$\therefore \quad \text{Re} = \frac{0.08 \times 0.4 \times 4}{0.08^2 \times \pi \times 3.57 \times 10^{-4}} = 17\,820$$

The water flow is therefore turbulent, and Eq (Int.16) may be used to determine the surface coefficient of heat transfer as follows:

$$h_c = \frac{0.023 \times 0.668 \times 17\,820^{0.8} \times 2.22^{0.33}}{0.08}$$

$$= 628.7 \text{ W/m}^2\,^\circ\text{C}$$

The changes in temperature may be shown as follows:

Pipe surface temperature	100°C	100°C
Water temperature	50°C $\longrightarrow$	70°C
	Δt_{max} 50°C Δt_{min}	30°C

From Eq (Int.24)

$$\Delta t_m = \frac{50 - 30}{\log_e \dfrac{50}{30}} = 39.1\,^\circ\text{C}$$

It is interesting to note that the arithmetic mean temperature difference would be: $100 - \frac{1}{2}(50 + 70) = 40\,^\circ\text{C}$, which is very near to the true mean of 39.1°C. In general, the arithmetic mean may be used instead of the logarithmic mean in cases where $\Delta t_{max} < 1.5\,\Delta t_{min}$. In this example $\Delta t_{max} = 1.5\,\Delta t_{min}$.

Since the heat gained by the water must equal the heat transferred at the surface of the tube, we have:

$$0.4 \times 4\,196(70 - 50) = \pi \times 0.08 \times l \times 628.7 \times 39.1$$

from which
$$l = 5.43 \text{ m}$$

Example Int.5. A freely exposed 3 m $\times$ 2 m $\times$ 2 m deep boiler feed tank contains water at 90°C. If the ambient air temperature is 20°C and the mean temperature of the surrounding surfaces is 15°C calculate:

    (*a*) the thickness of plastic insulating material having a thermal conductivity of 0.12 W/m°C which should be applied to the sides and lid of the tank if the external surface of the insulation is not to exceed 40°C,

    (*b*) the rate of heat loss from the sides and lid of the tank when (i) insulated and (ii) uninsulated.

    (*a*) Neglect the thickness and thermal resistance of the tank and assume that the external surface of the tank is uniformly at 90°C, then from Eq (Int.4) the heat conducted across the insulation will be:

$$\frac{\Phi}{A} = \frac{0.12}{l}(90 - 40)$$

$$= \frac{6}{l} \text{ W/m}^2 \qquad\qquad (1)$$

Heat will be lost by natural convection and by radiation from the sides and lid of the tank. Consider first the sides of the tank:

Heat emission by natural convection:

Using Eq (Int.23) and taking C = 1.9 for the vertical sides of the tank we have:

$$\frac{\Phi_c}{A} = 1.9(40 - 20)^{1.25}$$

$$= 80.35 \text{ W/m}^2$$

Heat emission by radiation:

Using Eq (Int.31) and Case 3 of Table Int.4 and assuming that $\epsilon = 0.9$ we have for $F_A = 1.0$

$$\frac{\Phi_r}{A} = 5.67 \times 0.9 \left[\left(\frac{40 + 273}{100}\right)^4 - \left(\frac{15 + 273}{100}\right)^4 \right]$$

$$= 138.6 \text{ W/m}^2$$

The total emission from the sides of the tank will be:

$$\frac{\Phi}{A} = \frac{\Phi_c}{A} + \frac{\Phi_r}{A} = 80.35 + 138.6$$

$$= 218.95 \text{ W/m}^2 \tag{2}$$

This must be equal to the heat conducted across the insulation, i.e. from (1) and (2)

$$\frac{6}{l} = 218.95$$

from which $l = 28$ mm

Consider now the lid of the tank. The heat conducted across the insulation will be the same as (1) above, and the heat emission by radiation will be the same as for the sides. The natural convection will, however, be different. Using Eq (Int.23) and taking C = 2.5 for heat flow upwards from the horizontal lid, we have:

$$\frac{\Phi_c}{A} = 2.5(40 - 20)^{1.25}$$

$$= 105.7 \text{ W/m}^2$$

The total emission from the surface of the lid will therefore be:

$$105.7 + 138.6 = 244.3 \text{ W/m}^2 \tag{3}$$

Then from (1) and (3)

$$\frac{6}{l} = 244.3$$

from which $l = 25$ mm

(*b*) (i) The heat emission from the insulated tank will be from (2) and (3):

Sides = 218.95 × 2 × 2(3 + 2) = 4 379
Lid   = 244.3 × 3 × 2   = 1 465.8

Total emission = 5 844.8 W, say 5.85 kW

(ii) The heat emission by natural convection from the uninsulated tank will be:

$$\text{Sides:} \quad \frac{\Phi_c}{A} = 1.9(90 - 20)^{1.25} = 384.8 \text{ W/m}^2$$

$$\text{Lid:} \quad \frac{\Phi_c}{A} = 2.5(90 - 20)^{1.25} = 506.2 \text{ W/m}^2$$

The emission by radiation will be:

$$\frac{\Phi_r}{A} = 5.67 \times 0.9 \left[\left(\frac{90 + 273}{100} \right)^4 - \left(\frac{15 + 273}{100} \right)^4 \right]$$
$$= 535 \text{ W/m}^2$$

The heat emission from the uninsulated tank will therefore be:

Sides = (384.8 + 535) × 2 × 2(3 + 2) = 18 396
Lid   = (506.2 + 535) × 3 × 2   = 6 247.2

24 643.2 W, say 24.6 kW

The reduction in heat loss due to the insulation is:

$$24.6 - 5.85 = 18.75 \text{ kW}$$

i.e. $\dfrac{18.75}{24.6} \times 100 = 76\%$ saving

N.B. Further heat-transfer examples will be found in Chapters 1 and 2.

Problems

1. The exhaust gases from a gas turbine are used to heat 12.6 kg/s of water from 105°C to 170°C in a parallel-flow heat exchanger. If the hot gases enter the heat exchanger at 370°C and leave at 205°C and the total heating surface area is 930 m², calculate the overall heat transfer coefficient. Take the specific heat capacity of the water as 4 280 J/kg°C.
   Ans.: 33.29 W/m²°C.

2. A pipe having an inside diameter of 20 mm is maintained uniformly at 121°C. What length of pipe is needed to heat water flowing through it at 0.19 kg/s from 66°C to 77°C? Data: For water at 71.5°C.

$c = 4\,187 \text{ J/kg}^{\circ}\text{C}$

$\eta = 4.05 \times 10^{-4} \text{ Ns/m}^2$

$\lambda = 0.659 \text{ W/m}^{\circ}\text{C}$

Ans.: 74 cm (nearest)

3. Calculate the rate of heat loss by natural convection from a 38 mm external diameter horizontal pipe if its surface temperature is 93.3°C and the ambient air temperature is 15.6°C.

Ans.: 82 W/m

N.B. Further problems involving the use of heat-transfer principles will be found in Chapters 1 and 2.

1: The Heat Requirements of Heated Buildings

Introduction

The heat transmission through a plane solid material may be calculated from a knowledge of its conductance and surface temperatures. When considering the flow of heat through the fabric of a heated building, however, the surface temperatures are not normally known. It is necessary therefore to consider an overall coefficient of heat transmission which includes the conductance at the inside and outside surfaces and which may be used in conjunction with the air-to-air temperature difference. This overall coefficient of heat transmission or simply the thermal transmittance (U) is a function of the thermal conductivity and thickness of the material and of the surface conductances, but since the conductance at the external surface is influenced by weather conditions, it can have no precise value. Thermal transmittance values have been determined experimentally for numerous common building materials and traditional constructions and are readily available, but for new materials and composite structures it may be convenient to calculate the U value using standard values of thermal conductivity and surface conductance.

Fig. 1.1 shows diagrammatically how heat is transferred from inside a heated building through an external wall to the outside. The inside surface of the wall receives heat by convection from the warmer room air and by thermal radiation from warmer surfaces within the room such as internal walls and heated areas such as radiators. This heat is transferred by conduction to the external surface, where heat is given off by convection to the cooler external air and by thermal radiation to the cooler surfaces that the wall "sees". The U value of a given wall therefore depends also upon the number of internal and external wall surfaces which it "sees".

Assuming steady state, the amount of heat entering the inside surface will be:

$$\frac{\Phi}{A} = h_{ri}(t_{mi} - t_{si}) + h_{ci}(t_i - t_{si}) \qquad (1.1)$$

where

Φ = rate of heat flow, W
A = surface area, m^2

24

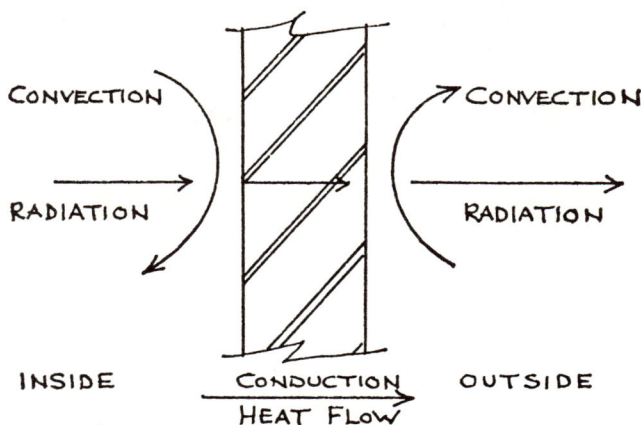

Fig. 1.1. Heat flow through an external wall.

t_{si} = surface temperature inside, °C
t_{mi} = mean radiant temperature of the surfaces seen by the inside surface at t_{si}, °C
t_i = inside air temperature, °C
h_{ri} = coefficient of heat transfer by thermal radiation at the inside surface, W/m²°C
h_{ci} = coefficient of heat transfer by convection at the inside surface,* W/m²°C

Or, alternatively,

$$\frac{\Phi}{A} = h_{si}(t_i - t_{si}) \qquad (1.2)$$

where h_{si} = a combined coefficient of heat transfer at the inside surface, W/m²°C.
From Eq (1.1) and Eq (1.2)

$$h_{si} = h_{ci} + \frac{h_{ri}(t_{mi} - t_{si})}{t_i - t_{si}} \qquad (1.3)$$

An alternative to Eq (1.2) would be:

$$\frac{\Phi}{A} = h_{si}'(t_i - t_{im}) \qquad (1.4)$$

where

$$h_{si}' = h_{ri} + h_{ci} \qquad (1.5)$$

* see later

and t_{im} = weighted mean of the air temperature t_i and the mean radiant temperature t_{mi}.

i.e.
$$t_{im} = \frac{h_{ri} \cdot t_{mi} + h_{ci} \cdot t_i}{h_{ri} + h_{ci}}$$
(1.6)

It is clear from the above that the surface conductance will vary from case to case. Methods of determining h_r and h_c are given in the first chapter "Introduction to Elementary Heat Transfer".

Calculation of Overall Coefficient of Heat Transmission

Fig. 1.2 shows a typical temperature gradient through a plane solid wall of constant cross-section. Heat flows from room air at a temperature t_i through the solid wall by conduction to outside air at a temperature t_o, the inside and outside surface temperatures being t_{si} and t_{so} respectively.

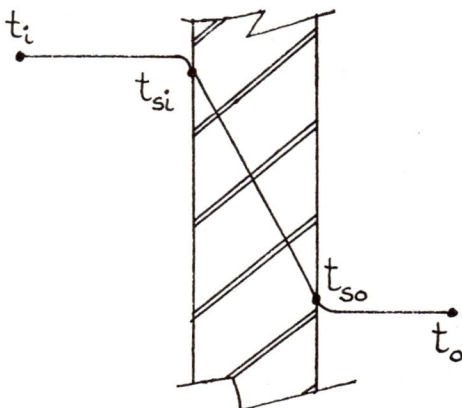

Fig. 1.2. Temperature gradient through a plane solid wall.

The flow of heat through unit area will be:

At inside surface:

$$\frac{\Phi}{A} = h_{si}(t_i - t_{si})$$
(1.7)

Through solid wall, applying Eq (Int.1.4):

$$\frac{\Phi}{A} = \frac{\lambda}{l}(t_{si} - t_{so})$$
(1.8)

At outside surface:

$$\frac{\Phi}{A} = h_{so}(t_{so} - t_o) \tag{1.9}$$

where subscript o refers to the outside.

Using the thermal transmittance U:

$$\frac{\Phi}{A} = U(t_i - t_o) \tag{1.10}$$

From Eq (1.7, 1.8, 1.9 and 1.10)

$$\frac{\Phi}{A} = U(t_i - t_o) = h_{si}(t_i - t_{si}) = \frac{\lambda}{l}(t_{si} - t_{so}) = h_{so}(t_{so} - t_o)$$

Transposing:

$$\frac{U}{h_{si}}(t_i - t_o) = t_i - t_{si}$$

$$\frac{Ul}{\lambda}(t_i - t_o) = t_{si} - t_{so}$$

$$\frac{U}{h_{so}}(t_i - t_o) = t_{so} - t_o$$

By addition

$$U\left(\frac{1}{h_{si}} + \frac{l}{\lambda} + \frac{1}{h_{so}}\right)(t_i - t_o) = t_i - t_o$$

from which

$$U = \left(\frac{1}{h_{si}} + \frac{l}{\lambda} + \frac{1}{h_{so}}\right)^{-1} \tag{1.11}$$

where

U = overall thermal transmittance, $W/m^2\,^\circ C$

h_{si} and h_{so} = conductance at inside and outside surfaces respectively, $W/m^2\,^\circ C$

λ = thermal conductivity of solid material, $W/m^\circ C$

l = thickness of solid material, m

It should be noted that h_{si}, h_{so} and $\frac{\lambda}{l}$ are conductances in series, and as it is usual to consider the flow of heat as analogous to the flow of electricity, their reciprocals $1/h_{si}$, $1/h_{so}$ and l/λ represent resistances to the flow of heat.

For a structure composed of a number of solid layers of thicknesses $l_1, l_2, \ldots l_n$ and thermal conductivities $\lambda_1, \lambda_2, \ldots \lambda_n$, and including an air space having a conductance of h_a, Eq (1.11) becomes

$$U = \cfrac{1}{\cfrac{1}{h_{si}} + \cfrac{l_1}{\lambda_1} + \cfrac{l_2}{\lambda_2} + \ldots \cfrac{l_n}{\lambda_n} + \cfrac{1}{h_a} + \cfrac{1}{h_{so}}}$$

or

$$U = \frac{1}{R_{si} + R_1 + R_2 + \ldots R_n + R_a + R_{so}} = \frac{1}{\Sigma R} \qquad (1.12)$$

where

R_{si} and R_{so} = resistances at inside and outside surfaces respectively, $m^2{}^\circ C/W$

R_a = resistance of air space, $m^2{}^\circ C/W$

$R_1, R_2, \ldots R_n$ = resistance of solid layers $1, 2, \ldots n$, $m^2{}^\circ C/W$

In cases where hollow blocks are used in the structure a further resistance term $R_h = \frac{1}{C}$, where C = measured conductance through the hollow block, should be included in the denominator of Eq (1.12).

The rate of heat loss per unit area of material when inside and outside air temperatures are both steady is, in general terms,

$$\frac{\Phi}{A} = \frac{\Delta t}{\Sigma R} \ W/m^2 \qquad (1.13)$$

where Δt = difference in temperature between inside and outside air, i.e. $t_i - t_o$

$\Sigma R = R_{si} + R_1 + R_2 + \ldots R_n + R_a + R_h + R_{so}$

Typical values of R_{si} and R_{so} are given in Tables 1.1 and 1.2 respectively.

TABLE 1.1.

Thermal resistance at inside plane* surfaces (R_{si})
$m^2{}^\circ C/W$

Surface	R_{si}
Wall (Emissivity = 0.9)	0.13
Ceiling or roof, heat flow upwards	0.11
do do downwards	0.14
Floor do upwards	0.11
do do downwards	0.14

* The resistance of a corrugated surface is about 80 per cent of the plane surface resistance.

TABLE 1.2.

Thermal resistance at outside plane surfaces (R_{so})
$m^2 {}^{\circ}C/W$

Surface	Exposure*		
	Sheltered	*Normal*	*Severe*
	R_{so}		
Wall (Emissivity = 0.9)	0.08	0.05	0.03
Roof (do)	0.07	0.04	0.02

* Based on wind speed of 1.1, 3.7 and 8.9 m/s respectively in the case of the roof and 0.7, 2.5 and 5.9 m/s respectively in the case of the wall.

In the case of solid ground floors the rate of heat flow downwards depends on the plan aspect ratio of the floor, its size, and the number of exposed edges. Eq 1.12 cannot therefore be used. Typical measured U values for solid ground floors are given in Table 1.3.

TABLE 1.3.

U value of solid ground floors $W/m^2 {}^{\circ}C$ of inside/outside temperature difference

Floor dimensions	Four exposed edges	Two exposed edges at right angles
170 m x 65 m	0.11	0.06
65 m x 65 m	0.15	0.08
65 m x 16 m	0.32	0.18
30 m x 16 m	0.36	0.21
16 m x 16 m	0.45	0.25
8 m x 8 m	0.76	0.45

The approximate heat requirements of buildings expressed in W/m^3 of heated space are given in Table 1.4.

TABLE 1.4.

Approximate heat requirements of buildings
W/m^3 of heated space

Dwellings	40 to 60
Buildings up to 3000 m^3	30 to 40
Buildings above 3000 m^3	15 to 30

Regulations under the Thermal Insulation (Industrial Buildings) Act, 1957 require that the sum of the surface resistances for factory roofs should be taken as 0.15 m²°C/W and that the U value shall not exceed 1.7 W/m²°C. If the design temperature of the building is lower than 21.1°C, or if parts of it are unheated, the standard will be satisfied if the rate of heat loss through the roof is not greater than 37.8 W/m² when the outside temperature is −1.1°C. The Building Regulations 1965 also specify that the sum of the surface resistances should be 0.15 m²°C/W for roofs and 0.18 m²°C/W for external walls.

The airspaces in hollow and lined walls, roofs and floors add considerable thermal resistance to the structure and therefore reduce the rate of heat loss from the building. Heat is transferred across the airspace by thermal radiation and also by convection the rate of heat flow depending on the emissivity of the surfaces in the case of radiation and on the thickness, ventilation rate and mean temperature in the case of convection. For most ordinary building materials the emissivity is high, about 0.9, and if both surfaces of a cavity have this emissivity about two-thirds of the total heat flow is by radiation. If the airspace is lined with reflective insulation, say aluminium foil, the radiation is considerably reduced particularly if multiple layer foil is used. If only one face of the airspace is lined with reflective material it does not matter on which face the insulation is placed for on the hot face it would act as a poor emitter and on the cold face as a good reflector. The effect of reflective insulation is illustrated in Problem 6. Convection across the airspace involves air movement and if the gap is less than about 20 mm there is considerable interference between the ascending and descending air streams resulting in greater turbulence and a corresponding increase in surface conductance. In general the resistance of an airspace decreases rapidly with decrease in thickness of the gap; for thicknesses about 20 mm the resistance is substantially constant. In the case of horizontal airspaces the resistance is greater for heat flow downwards than for heat flow upwards because downward convection is comparatively small. This is a particularly important consideration when determining heat gains through roofs in summer. Corrugated surfaces have a greater surface for convection heat transfer resulting in an increase in heat flow up to about 10 per cent depending also on the emissivity of the surfaces. Ventilation occurs in normal hollow wall construction and also in badly constructed lined structures resulting in an increase in the convection heat transfer and a lowering of the total resistance of the airspace. Table (1.5) gives the standard resistance (R_a) of some typical airspaces and shows the effect of reflective linings, thickness and ventilation.

It should be noted that the use of thermal insulation in buildings generally not only reduces the rate of heat loss but also: (i) increases the inside surface temperature and therefore helps to provide better conditions

TABLE 1.5.

Airspace resistance (R_a)
$m^2\,°C/W$

Ventilated (minimum thickness 20 mm)

Between asbestos cement or black metal cladding, unsealed joints and high emissivity lining.	0.16
ditto and low emissivity lining	0.3
Hollow wall construction, high emissivity surfaces	0.18
Tiles on tile-hung wall	0.12

Unventilated

Plane and corrugated sheets in contact, high emissivity surfaces:

Heat flow horizontal or upwards	0.09
Heat flow downwards	0.11

Multiple foil insulation, low emissivity surfaces:

Heat flow horizontal or upwards	0.58
Heat flow downwards	1.76

for comfort, (ii) brings about a change in the thermal capacity and hence thermal response of the structure and (iii) alleviates the risk of surface condensation. These features are dealt with in the illustrative examples that follow.

Example 1.1. List and discuss in general terms the factors that should be taken into account when determining the heat requirements of a heated building.

The heat requirements of a heated building are determined by several factors; which include:

(*a*) The design inside and outside temperatures.
(*b*) The degree of exposure of the building.
(*c*) The nature and thickness of the various building materials used in the construction.
(*d*) The rate of air change for ventilation.
(*e*) The method of heating, temperature gradients and the corresponding allowance to be made for height of heated spaces.
(*f*) The allowance for intermittent heating.
(*g*) The allowance for any heat gains.
(*h*) Any margins.

(a) *Inside and Outside Temperatures*. Since the heating installation must be capable of providing adequate warmth and comfort during the coldest period in winter, the designer is concerned mainly with the maximum heat requirement of the building.

While the warmth of an environment depends upon air temperature, mean radiant temperature, air humidity and air movement, it is usual when simple heating appliances are used to consider the air and mean radiant temperatures only in the form of an environmental temperature (t_e) and in Eq 1.10 the difference between the inside and outside environmental temperatures may be used, i.e.

$$\frac{\Phi}{A} = U(t_{ei} - t_{eo})$$

where

t_{ei} = a weighted mean between the mean surface temperature (t_{mi}) seen by the heat losing surface and the room air temperature (t_i).
t_{eo} = the outside sol-air temperature.

In most cases $t_{ei} = \frac{2}{3} t_{mi} + \frac{1}{3} t_i$ but since t_{mi} is not normally known at the outset of a design and must be assumed and because during winter the outside sol-air temperature is approximately equal to the outside air temperature, the use of $t_i - t_o$ may be more convenient when determining winter heat losses.

Tables of recommended environmental temperatures for all classes of buildings are given in *A Guide to Current Practice 1970*, published by the Institution of Heating and Ventilating Engineers.

Basic design external air temperatures for space heating are selected generally in accordance with the meteorological data given in Post-war Building Study No. 33 and Section 1 of the 1HVE Guide 1970. For multi-storey buildings with solid intermediate floors and partitions, and assuming a system overload capacity of 20 per cent, the external design temperature is $-1°C$. If there is to be no system overload capacity, then $-4°C$ should be used. Similar figures for single-storey buildings are $-3°C$ and $-5°C$ respectively.

(b) *Exposure.* The rate of heat loss from the external surfaces of a heated building depends upon the severity of its exposure to winds. Three degrees of exposure are generally assumed, they are:

"Sheltered"—includes the first two storeys above ground of buildings in towns.
"Normal"—includes the third, fourth and fifth storeys of buildings in towns and most suburban and country premises.
"Severe"—includes sixth and higher storeys of buildings in towns and buildings on hill sites, the coast or river-side.

If the rate of heat loss from a building having a normal exposure is taken as a basis for comparison the rate of heat loss for a sheltered exposure may be from 5 to 10 per cent less and for a severe exposure from 10 to 15 per cent more.

(c) *Nature and Thickness of the Building Materials.* The rate of heat loss through the fabric of a building depends upon the materials used, their thickness and thermal conductivity, the temperature difference between inside and outside air, the degree of exposure and upon the value of the thermal resistance at the inside surface. Ref Eq (1.11), (1.12) and (1.13).

(d) *Air Change.* The air change for ventilation may be provided by natural infiltration or by mechanical means, depending upon the use and size of the room and the number of occupants. Certain minimum standards of ventilation are laid down for most classes of buildings, and for typical values reference should be made to the *Guide to Current Practice 1970* of the Institution of Heating and Ventilating Engineers. The heating installation must provide sufficient heat to warm the cold air entering the building for ventilation purposes, and the amount of heat required may be calculated as shown in Example 1.9.

Uncontrolled air change leads to fuel wastage and is dependent on many factors such as: the number of external walls; height of the building; the tightness of the building construction and the number and type of windows and doors; the number and position of staircases, lift shafts and corridors; the pressure gradient across the building due to the direction and strength of the wind; inside/outside temperature gradient and variations in occupancy. Since the heating installation must provide sufficient heat to warm the cold air entering the building for ventilation purposes it is important that the building should be as airtight as possible and that uncontrolled air change is reduced to a minimum.

(e) *Method of Heating.* Space heating appliances emit heat either primarily by convection or primarily by radiation. Convection heaters generally cause greater temperature gradients than radiant heaters, and to allow for this an allowance for the height of the heated space is made to the net heat loss based on the design inside/outside air temperature difference. The values in Table 1.6 are taken from the *Guide to Current Practice* of the Institution of Heating and Ventilating Engineers. For lofty spaces, such as churches, where heating surfaces are installed at the level of the clerestory windows, the heated space should be regarded as two spaces, one from the floor to the level of the upper heating surfaces, and the other from the upper heating surfaces to the roof.

(f) *Intermittent Heating.* While heating systems are normally designed to meet the steady heat loss determined from the basic design inside/outside temperature difference steady conditions rarely, if ever, prevail. Most buildings are heated intermittently and therefore undergo regular periods of heating up and cooling down the rate of heating and cooling depending largely on the thermal capacity, i.e. heat storage of the structure. Buildings which have a low thermal capacity respond rapidly to changing conditions whereas buildings with a high thermal capacity respond slowly. The time lag depends also on the ventilation rate, the incidence of direct solar radiation, the thermal capacity of the heating installation and contents of

TABLE 1.6

Allowance for Height of Heated Space

Percentage addition to net heat loss

Method of heating	Height of space	
	6 m	10 m
Natural convection: radiators, convectors, skirting heaters and similar appliances	2	5
Forced warm air: cross flow from low level inlets	5	12
downward flow from high level inlets	3	8
Floor heating	nil	nil
Ceiling heating	3	10
Radiant panels at high level	nil —	2

the building and on the changes that take place in the outside temperature. As far as the structure is concerned, say for example an external wall, the thermal diffusivity, $a = \lambda/\rho c$, gives a good indication of the rate at which a temperature change is propogated through the material; the rate increasing as the value of a increases. It may also be shown that the wall will heat up rapidly if the rate of heat input (Φ) is large and the property $\lambda\rho c$ is small. An analytical discussion of unsteady heat flow is outside the scope of this book and the reader is referred to the work of Billington* on this subject. The heat storage capacity of the structure gives some idea of the response of a building to changing conditions, see Example 1.2.

The time lag of heating systems depends on their thermal capacity. Low thermal capacity systems such as direct gas and electric radiant panels and warm air systems are suitable for use in low thermal capacity buildings whereas systems having a large thermal capacity such as hot water systems and embedded pipes and cables are more suitable for use in high thermal capacity buildings. Thermal storage systems using off-peak electricity, such as block storage heaters and embedded cables, are suitable only if the building has a high thermal capacity.

The many variables preclude a simple solution to intermittent heating, but it is known from experience that less fuel is required than for continuous heating and that standards may be maintained provided that the heating plant has a reserve of power to re-heat the system and building after a period of shut down. Fuel economies are greatest for 'light' buildings with low thermal capacity systems having a large reserve power and least for 'heavy' buildings with high thermal capacity systems with small reserve power. The length of the re-heat period is determined by the available reserve power and will be considerably shorter during mild weather than during cold spells. At times when the external temperature is below the design value the reserve power would be required to maintain internal

* Billington, N. S. Building Physics: Heat 1967, Pergamon Press.

standards even with continuous heating. As it is not possible to develop an easy method of determining the optimum size of heating plant we must rely on experience which suggests the margins given in Table 1.7.

TABLE 1.7

Suggested Margins for Intermittent Heating

Percentage addition to be made to net total heat loss, i.e. fabric + air change.

Heating method	Period of re-heat	Period of occupation 7 day/week	5 day/week
1. Hot water radiators, convectors, skirting heaters and pipe coils........	3 4	25 20	30 25
2. Steam radiators, convectors and pipe coils	3 4	20 15	25 20
3. Unit heaters, industrial radiant panels and warm air in factory type buildings	2	45	50
4. Direct electric, gas and oil heaters	3	25	30

Notes 1. In the case of steam systems operating at constant pressure the margin should be added to the heating surface as well as to the total heat loss.
2. In the case of item No. 4 additional heating surface must also be provided.
3. In cases where the thermal capacity of the building is very low the margin may be reduced by no more than 25 per cent.

(g) *Heat gains.* Internal heat gains from occupants, lighting, motors and machinery and solar-heat gains should be allowed only if they are appreciable and fairly continuous. While in many instances these sources of heat do not affect the size of the heating installation, they may have a significant effect on its operation and control and on the fuel consumption.

Example 1.2. (*a*) Derive an expression for the thermal capacity of insulated and uninsulated walls, (*b*) Write notes on the effect of thermal capacity making particular reference to the position of thermal insulation.

Consider Fig. 1.3 which shows the steady heat flow temperature gradient through an uninsulated plane solid wall. Applying the resistance concept introduced earlier we have:

$$\frac{\Phi}{A} = \frac{t_m - t_o}{\frac{l}{2\lambda} + R_{so}}$$

which for steady heat flow must be equal to $U(t_i - t_o)$, i.e.

$$t_m - t_o = U(t_i - t_o)\left(\frac{l}{2\lambda} + R_{so}\right)$$

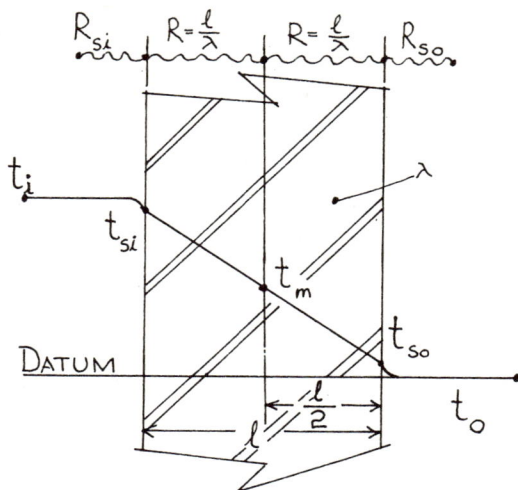

Fig. 3.

and the thermal capacity of the wall expressed per unit area and measured above t_o will be:

$$Q = m c \, U(t_i - t_o)\left(\frac{l}{2\lambda} + R_{so}\right) \tag{1.14}$$

where m = mass per unit area of wall surface,
c = specific heat capacity of wall material.

If alternatively the thermal capacity is expressed per degree difference between inside and outside temperatures, we have:

$$Q = m c \, U\left(\frac{l}{2\lambda} + R_{so}\right) \tag{1.15}$$

For an insulated wall having two components 1 and 2, Fig. 1.4, the thermal capacity will be:

$$Q = Q_1 + Q_2$$

where

$$Q_1 = m_1 c_1 \, U\left(\frac{l_1}{2\lambda_1} + R_2 + R_{so}\right) \tag{1.16}$$

and

$$Q_2 = m_2 c_2 \, U\left(\frac{l_2}{2\lambda_2} + R_{so}\right) \tag{1.17}$$

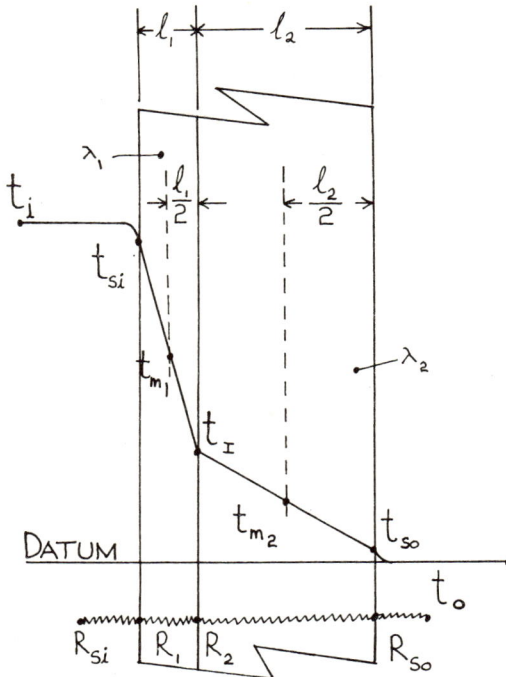

Fig. 1.4

$$t_{m1} = t_o + U(t_i - t_o)\left(\frac{l_1}{2\lambda_1} + R_2 + R_{so}\right)$$

$$t_{m2} = t_o + U(t_i - t_o)\left(\frac{l_2}{2\lambda_2} + R_{so}\right)$$

$$t_{si} = t_i - U(t_i - t_o)R_{si}$$

$$t_{so} = t_o + U(t_i - t_o)R_{so}$$

$$t_I = t_i - U(t_i - t_o)(R_{si} + R_1)$$

It should be noted that the total thermal capacity will have a maximum value when the material which offers the greatest resistance to heat flow is on the outside and a minimum when on the inside. This is shown in Table 1.8 which refers to the three walls shown in Fig. 1.5. Wall A is uninsulated, Wall B has insulation on the inside and Wall C has insulation on the outside. The relevant data for the wall and insulation are as follows:—

$$t_i = 21.1°C$$
$$t_o = -1.1°C$$

$R_{si} = 0.12 \text{ m}^2{}^\circ\text{C/W}$
$R_{so} = 0.05 \text{ m}^2{}^\circ\text{C/W}$
  $l = 25.4$ cm, wall; 7.6 cm, insulation
  $\lambda = 1.44 \text{ W/m}^\circ\text{C}$, wall; $0.144 \text{ W/m}^\circ\text{C}$, insulation
  $\rho = 2\,400 \text{ kg/m}^3$, wall; 641 kg/m^3, insulation
  $c = 837 \text{ J/kg}^\circ\text{C}$, wall; $837 \text{ J/kg}^\circ\text{C}$, insulation

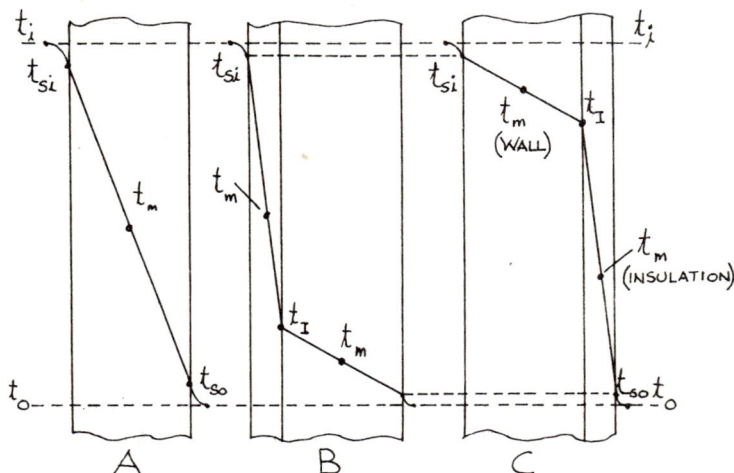

Fig. 1.5

TABLE 1.8

Comparative Data for the Walls in Fig. 1.5

Wall		A	B	C
U	W/m^2 $^\circ$C	2.84	1.14	1.14
t_{si}	$^\circ$C	13.3	18.0	18.0
t_{so}	$^\circ$C	2.2	0.22	0.22
(t_m) wall	$^\circ$C	7.8	2.44	2.44
(t_m) insulation	$^\circ$C	–	11.33	6.89
t_I	$^\circ$C	–	4.67	13.56
Q : wall	kJ/m^2 $^\circ$C	204.3	81.7	388.2
Q : insulation	kJ/m^2 $^\circ$C	–	22.9	14.7
Q : total	kJ/m^2 $^\circ$C	204.3	104.6	402.9
Q/U	h	20	25.6	98.6

Table 1.8 shows that $U_A > U_B = U_C$ and that $Q_A > Q_B < Q_C$. Using wall A as reference, the saving in thermal transmittance when insulation is used is 60 per cent. The reduction in thermal capacity in wall B is 48.8 per

cent and the increase in wall C is 97.2 per cent. While wall B has the same thermal transmittance as wall C it has a much lower thermal capacity and would therefore respond more rapidly to temperature changes than wall C. In general, buildings which have insulation on the inside are more suitable for intermittent use and heating than those which have insulation on the outside. Buildings having a high thermal capacity are generally more suitable for continuous use and may be heated by thermal storage methods. The ideal for overall economy would be a low thermal capacity heating system in a low thermal capacity building. Table 1.8 also gives the thermal time constant Q/U for each of the three walls. It may be shown that the time lag is dependent on this constant and that in general slow cooling will take place only if Q/U is comparatively large.

Example 1.3. Calculate and compare the coefficients of thermal transmittance of a 220 mm thick solid brick wall and a 260 mm thick cavity brick wall.

Thermal conductivity of brick	1.15 W/m°C
Inside surface resistance	0.123 m²°C/W
Outside surface resistance	0.053 m²°C/W
Air space resistance	0.176 m²°C/W

For the 220 mm thick solid brick wall, from Eq (1.11),

$$U = (0.123 + \frac{0.220}{1.15} + 0.053)^{-1} = 2.7 \text{ W/m}^2\text{°C}$$

For the 260 mm thick cavity brick wall (i.e. 105 mm brick; 50 mm air space; 105 mm brick), from Eq (1.12),

$$U = (0.123 + \frac{0.210}{1.15} + 0.176 + 0.053)^{-1} = 1.87 \text{ W/m}^2\text{°C}$$

The saving in thermal transmittance when a 260 mm thick cavity wall is used instead of a 220 mm thick solid brick wall is

$$\frac{2.7 - 1.87}{2.7} \times 100 = 31 \text{ per cent}$$

Example 1.4. A closed roof space having a pitched roof and a heated room below is shown in Fig. 1.6. Neglecting the effect of heat loss through the end walls, derive the expression

$$U = \frac{U_r \times U_c}{U_r + U_c \times \cos \theta}$$

where

U = combined thermal transmittance coefficient for the ceiling and roof
U_r = thermal transmittance for roof
U_c = thermal transmittance for ceiling
θ = angle of pitched roof

Let A_c = area of ceiling
A_r = area of roof
t_1 = temperature of air in room below
t_2 = temperature of air in closed roof space
t_3 = temperature of outside air.

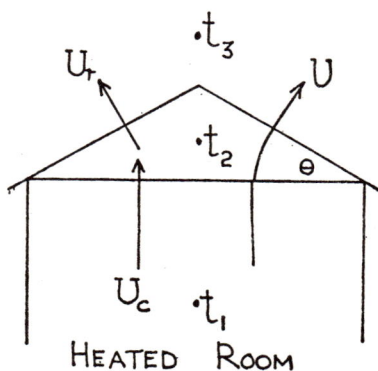

Fig. 1.6. Closed roof space

Then for steady heat flow conditions and since $A_r = A_c/\cos \theta$ the heat loss will be

$$U_c \cdot A_c(t_1 - t_2) \tag{1.18}$$

or

$$U_r \cdot \frac{A_c}{\cos \theta}(t_2 - t_3) \tag{1.19}$$

or

$$U \cdot A_c(t_1 - t_3) \tag{1.20}$$

From Eqs (1.18) and (1.19)

$$\frac{U_c}{U_r} \cdot \cos \theta(t_1 - t_2) = t_2 - t_3 \tag{1.21}$$

From Eqs (1.18) and (1.20)

$$\frac{U_c}{U}(t_1 - t_2) = t_1 - t_3 \tag{1.22}$$

Subtract Eq (1.21) from Eq (1.22)

$$\frac{U_c}{U}(t_1 - t_2) - \frac{U_c}{U_r} \cdot \cos \theta(t_1 - t_2) = (t_1 - t_3) - (t_2 - t_3) = t_1 - t_2$$

Divide throughout by $t_1 - t_2$

$$\frac{U_c}{U} - \frac{U_c}{U_r} \cdot \cos \theta = 1$$

from which by transposing

$$U = \frac{U_c \times U_r}{U_r + U_c \cdot \cos \theta}$$

Example 1.5. The air in a room is maintained at 20°C when the outside air is at 0°C. Calculate the inside surface temperature when the wall construction is alternatively:

(*a*) 220 mm thick solid brickwork.

(*b*) 260 mm thick hollow brick wall having a 50 mm cavity and 12 mm thick plaster applied to the inside surface.

(*c*) 100 mm thick concrete with 50 mm studding on inside surface covered with 12 mm thick plasterboard to form an air space packed with glass fibre.

Data:

Thermal resistance at inside surface	0.13 m^2°C/W
Thermal resistance of air space	0.18 ” ”
Thermal resistance at outside surface	0.05 ” ”
Thermal conductivity of brick	1.15 W/m°C
Thermal conductivity of plaster	0.58 ”
Thermal conductivity of concrete	1.44 ”
Thermal conductivity of plasterboard	0.14 ”
Thermal conductivity of glass fibre	0.05 ”

In general, let

t_i = temperature of room air
t_{si} = inside surface temperature of wall
R_{si} = thermal resistance at inside surface

Then since heat flow is constant, from Eq (1.13)

$$\frac{\Phi}{A} = \frac{\Delta t}{\Sigma R} = \frac{t_i - t_{si}}{R_{si}}$$

from which

$$t_{si} = t_i - R_{si} \cdot \frac{\Delta t}{\Sigma R}$$

Then in this example

$$t_{si} = 20 - \frac{0.13(20)}{\Sigma R}$$

i.e.

$$t_{si} = 20 - \frac{2.6}{\Sigma R} \tag{1.23}$$

Total thermal resistance (ΣR) for each wall will be:

Wall (a) $\Sigma R = 0.13 + \dfrac{0.220}{1.15} + 0.05 \doteqdot 0.37 \text{ m}^2{}^\circ\text{C/W}$

" (b) $\Sigma R = 0.13 + \dfrac{0.210}{1.15} + 0.18 + \dfrac{0.012}{0.58} + 0.05 = 0.56 \text{ m}^2{}^\circ\text{C/W}$

" (c) $\Sigma R = 0.13 + \dfrac{0.1}{1.44} + \dfrac{0.05}{0.05} + \dfrac{0.012}{0.14} + 0.05 = 1.34 \text{ m}^2{}^\circ\text{C/W}$

Then inside surface temperature (t_{si}) for each wall will be from Eq (1.23).

Wall (a) $t_{si} = 20 - \dfrac{2.6}{0.37} = 13^\circ\text{C}$

" (b) $t_{si} = 20 - \dfrac{2.6}{0.56} = 16.4^\circ\text{C}$

" (c) $t_{si} = 20 - \dfrac{2.6}{1.34} = 18.1^\circ\text{C}$

Example 1.6. A room maintained at 20°C has a 220 mm thick brick external wall. Calculate the steady rate of heat loss from the external surface of the wall when exposed to air at 2°C and surroundings having a mean radiant temperature of −4°C.

Thermal conductivity of brick	1.15 W/m°C
Thermal resistance at inside surface	0.13 m²°C/W
Coefficient of heat transmission by radiation at external surface	4.5 W/m²°C
Coefficient of heat transmission by convection at external surface	3.4 W/m²°C

In previous examples involving the concept of thermal transmittance (U) it has been assumed that the air and surroundings are uniformly at the same temperature and a combined surface conductance for radiation and convection at the surface has therefore been used. With the present example sufficient data is given to enable the heat transfer from the external surface of the wall to be found from

$$\frac{\Phi}{A} = h_r(t_{s_o} - t_m) + h_c(t_{s_o} - t_o) \qquad (1.24)$$

where

Φ = total rate of heat transfer, W
h_r = coefficient of heat transfer by radiation, W/m²°C
h_c = coefficient of heat transfer by convection, W/m²°C
t_{s_o} = temperature of external surface of wall, °C

t_o = temperature of outside air, °C

t_m = mean radiant temperature of surroundings, °C

A = area, m²

For steady state, the rate of heat flow from inside air to the external wall surface will equal the total heat loss from the external surface by radiation and convection, i.e.:

$$\frac{20 - t_{S_o}}{0.13 + \frac{0.22}{1.15}} = 4.5(t_{S_o} + 4) + 3.4(t_{S_o} - 2)$$

from which

$$t_{S_o} = 4.3°C$$

The rate of heat loss from the external surface will be, from Eq (1.24),

$$\frac{\Phi}{A} = 4.5(4.3 + 4) + 3.4(4.3 - 2)$$

$$= 35.2 \text{ W/m}^2$$

Example 1.7. Calculate the outside temperature at which condensation will begin to occur on the inside surface of a 220 mm thick solid brick external wall if the room air dry bulb and dew point temperatures are 20°C and 13°C respectively.

Thermal conductivity of brick	= 1.15 W/m°C
Thermal resistance at inside surface	= 0.13 m²°C/W
Thermal resistance at outside surface	= 0.05 m²°C/W

The dewpoint temperature is that temperature to which the room air must be cooled in order to saturate it without increase in moisture content. Condensation will therefore begin to occur when the surface temperature of the wall is at or below the dewpoint temperature of 13°C when the rate of heat flow at the inside surface will be, from Eq (1.13),

$$\frac{\Phi}{A} = \frac{20 - 13}{0.13} = 53.9 \text{ W/m}^2$$

For steady state conditions this rate of heat flow is constant and equal to the rate of heat flow from the inside surface at 13°C to outside air at, say, t_o, i.e.

$$53.9 = \frac{13 - t_o}{\frac{0.22}{1.15} + 0.05}$$

from which

$$t_o = 0°C \text{ approx.}$$

Example 1.8. It is required to maintain the corner room of a single storey building at 20°C when 0°C outside.

(*a*) Using the data listed below, calculate the steady rate of heat loss from the room.

(*b*) Assuming that heating appliances are installed in the room to meet the above heat loss, what inside temperature could be maintained with the same heat input if the lining is removed?

Data:

Wall: lined with 15 mm thick fibreboard to form an air space;
        U = 1.31 W/m²°C; area = 27 m²
Roof: lined with 15 mm fibreboard to form an air space;
        U = 1.48 W/m²°C; area = 35 m²
Glass: U = 5.68 W/m²°C; area = 9 m²
Floor: U = 0.45 W/m²°C; area = 35 m²
Thermal conductivity of fibreboard = 0.06 W/m°C
Thermal resistance of air space = 0.18 m²°C/W
Air change rate = 0.05 m³/s   (NC)*

$$\text{Total resistance of lined wall} = \frac{1}{U} = \frac{1}{1.31} \text{ m}^2\,°\text{C/W}$$

When the fibreboard lining is removed the total resistance of the lined wall will be reduced by the resistance of the fibreboard lining and the air space. The thermal transmittance of the unlined wall is therefore:

$$U = \frac{1}{\dfrac{1}{1.31} - 0.18 - \dfrac{0.015}{0.06}} = 3.0 \text{ W/m}^2\,°\text{C}$$

Similarly, the thermal transmittance of the unlined roof will be

$$U = \frac{1}{\dfrac{1}{1.48} - 0.18 - \dfrac{0.015}{0.06}} = 4.1 \text{ W/m}^2\,°\text{C}$$

Taking the density of air at 1.2 kg/m³ and the specific heat capacity of air at 1.0 kJ/kg°C, the air change will require a heat input of $1.2 \times 1.0 \times 10^3 = 1\,200$ J/m³°C.

---

* Refers throughout to data from the National College for Heating, Ventilating, Refrigeration and Fan Engineering.

The total heat loss per degree difference between inside and outside air may now be calculated as follows:

Item	Vol. or area	U value or factor		Heat loss, W/°C	
		Lined	Unlined	Lined	Unlined
Air change ...	0.05 m³/s	1 200	1200	60	60
Glass	9 m²	5.68	5.68	51.12	51.12
Wall	27 "	1.31	3.0	35.37	81
Roof	35 "	1.48	4.1	51.8	143.5
Floor	35 "	0.45	0.45	15.75	15.75
			Totals =	214	351.4

Then for 20°C difference we have:

Heat loss for lined room = 214 × 20 = 4 280 W

Let t_x = inside air temperature of unlined room.
Then, since t_o = 0°C:

Heat loss for unlined room = 351.4 t_x

Since heat input is constant

$$351.4 \, t_x = 4\,280$$

$$t_x = \frac{4\,280}{351.4}$$

and $$t_x = 12.2°C$$

Example 1.9. The wall of a building is constructed from corrugated steel sheeting fixed to wooden studding and insulated inside by a fibreboard lining to form a sealed air space. The inner surface temperature of the insulated wall is not to fall below 8°C when the inside and outside air temperatures are 10°C and −1°C respectively. Using the data listed below, determine the thickness of the fibreboard insulating material.

Thermal transmittance of corrugated steel sheeting = 6.8 W/m²°C
Thermal resistance of air space = 0.18 m²°C/W
Thermal resistance at internal surface = 0.13 m²°C/W
Resistivity of fibreboard = 17.34 m°C/W

Rate of heat flow is given by Eq (1.13), and since inside surface temperature is given

$$\frac{\Phi}{A} = \frac{10 - 8}{0.13} = 15.38 \text{ W/m}^2$$

and required thermal transmittance for insulated wall will therefore be

$$\frac{15.38}{10 - (-1)} = 1.39 \text{ W/m}^2{}^\circ\text{C}$$

The total thermal resistance to heat flow through the corrugated steel sheeting alone is

$$\frac{1}{6.8} = 0.147 \text{ m}^2{}^\circ\text{C/W}$$

which includes the normal inside and outside surface resistances. For the insulated wall the total resistance must also include the resistance of the air space and the resistance of the fibreboard lining.

Let l = thickness of fibreboard lining then, since resistivity is the reciprocal of conductivity.

$$1.39 = \frac{1}{0.147 + 0.18 + 17.34l}$$

from which

$$l = 0.023 \text{ m} = 23 \text{ mm}$$

Example 1.10. An unheated store-room 2 m × 1.5 m × 3 m high has two external walls, and is separated from an office by two partitions. The long external wall has a thermal transmittance of 2.3 W/m²°C; the shorter external wall has a thermal transmittance of 1.7 W/m²°C. The two internal partitions are formed of 15 mm thick insulating board, plastered 15 mm thick and spaced 50 mm apart by framework. Compute the temperature of the store when the outdoor temperature is −2.8°C and the office temperature is 17°C. Neglect heat gains or losses through the floor or ceiling.

Air change rate	0.003 m³/s
Thermal conductivity of insulating board	0.05 W/m°C
Thermal conductivity of plasterboard	0.24 " " "
Thermal resistance of air space	0.18 m°C/W
Thermal resistance at internal surface	0.13 " " "
Specific heat of air	1200 J/m³°C

(NC)

Thermal transmittance of partition, from Eq (1.12),

$$U = \frac{1}{0.13 + 2\left(\frac{0.015}{0.24}\right) + 0.18 + 2\left(\frac{0.015}{0.05}\right) + 0.13}$$

$$= 0.86 \text{ W/m}^2{}^\circ\text{C}$$

Let t_x = air temperature in unheated store.
Then heat loss from store to outside will be:
Air change: 0.003 × 1 200 =   3.6
Long wall: 2 × 3 × 2.3   = 13.8
Short wall: 1.5 × 3 × 1.7 = 7.65

$$\overline{25.05} \text{ W/}^\circ\text{C}$$

$$= 25.05(t_x + 2.8)\text{W}$$

Heat gained by store from office will be:

Partitions: 3(2 + 1.5) × 0.86 = 9.03 W/$^\circ$C
$$= 9.03(17 - t_x)\text{W}$$

Assuming steady conditions, the heat lost from the store will equal the heat gained from the office, i.e.
$$25.05(t_x + 2.8) = 9.03(17 - t_x)$$
from which
$$t_x = 2.5^\circ\text{C approx.}$$

Example 1.11. Show that the thermal resistance of a sealed air space is approximately 0.18 m^2 $^\circ$C/W.

Heat is transferred across an air space partly by radiation from one surface to the other and partly by convection currents in the air itself. The radiation may be calculated from the Stefan and Boltzmann law:

$$\frac{\Phi_r}{A} = 5.67 \times 10^{-8} \text{ F}_E . \text{F}_A(T_1^4 - T_2^4) \text{ W/m}^2$$

where
F$_E$ = emissivity factor
F$_A$ = geometry factor
T$_1$ and T$_2$ = temperature of surfaces. K

Since an air space is bounded by two large parallel planes F$_A$ = 1.0 and
F$_E$ = $e_1 e_2/(e_1 + e_2 - e_1 e_2)$, where e = emissivity of surface.
For most building materials $e \geqslant 0.8$ and F$_E$ = 0.82.

The radiation may also be determined from

$$\frac{\Phi_r}{A} = h_r . \text{F}_E(t_1 - t_2) \text{ W/m}^2$$

where
$$h_r = 5.67 \times 10^{-8} \frac{(T_1^4 - T_2^4)}{t_1 - t_2} \text{ W/m}^2{}^\circ\text{C}$$

and t_1 and t_2 = temperature of surfaces $^\circ$C.
For $(t_1 - t_2) \leqslant 55^\circ$C, the mean value of h_r is found to be 5.0, then, taking F$_E$ as 0.82 for two brick surfaces:

$$\frac{\Phi_r}{A} = 5.0 \times 0.82(t_1 - t_2)$$
$$= 4.1(t_1 - t_2) \text{ W/m}^2$$

For air spaces between 20mm and 150mm in width the convection transfer may be calculated from:

$$\frac{\Phi_c}{A} = h_c(t_1 - t_a) \text{ W/m}^2$$

where t_1 = temperature of warm face, °C

t_a = mean temperature of air within space, i.e. $(t_1 + t_2)/2$, °C.

For a vertical air space with $(t_1 - t_a) \leqslant 55$°C the mean value of h_c is 2.84.

Then, substituting for t_a:

$$\frac{\Phi_c}{A} = 1.42(t_1 - t_2) \text{ W/m}^2$$

The total rate of heat transfer will then be:

$$\frac{\Phi}{A} = (4.1 + 1.42)(t_1 - t_2)$$

$$= 5.52(t_1 - t_2) \text{ W/m}^2$$

∴ Resistance of air space, $R_a = \dfrac{1}{5.52} = 0.18 \text{ m}^2{}°\text{C/W}$ approx.

Problems

1. The flat roof of a swimming-bath consists of concrete 152mm thick covered with asphalt 13mm thick. In order to avoid condensation, it is necessary that the temperature of the underside of the concrete should not fall below 15.6°C when the indoor and outdoor air temperatures are 18.3°C and 1.7°C respectively. What thickness of wood-wool slab insulation should be provided?

Thermal conductivity of concrete   = 1.44 W/m°C
Thermal conductivity of asphalt   = 1.15 " " "
Thermal conductivity of wood-wool   = 0.087 " " "
Thermal resistance at inside surface = 0.106 m²°C/W
Thermal resistance at external surface = 0.053 " " "

Ans.: 31mm approx.

2. The heat loss from a building is 0.293 MW for an inside temperature of 15.6°C and an outside temperature of −1.1°C. The window area in the building comprises 1 394 m² of single glazing having a thermal transmittance of 5.68 W/m²°C. If 70 per cent of the glass area is replaced with double glazing having a thermal transmittance of 2.84 W/m²°C, what maximum temperature will be attained in the building if the heat input, air change and outside temperature remain constant?   (NC)

Ans.: 19°C approx.

3. A factory roof consists of 6 mm-thick asbestos-cement sheeting having a thermal conductivity of 0.29 W/m°C. If the inside air temperature is 15.6°C and the dewpoint temperature is 10°C, calculate the outside air temperature at which condensation will start to occur on the inner surface of the roof.

Internal surface resistance = 0.106 m^2°C/W
External surface resistance = 0.044 m^2°C/W   (NC)

Ans.: 6.6°C approx.

4. A 152 mm-thick concrete external wall is insulated by a layer of fibreboard fixed to the inside surface by wooden studs. Using the data given, and assuming steady state conditions, determine:

(*a*) The thickness of fibreboard required to prevent the inside surface temperature of the insulated wall falling below 18.3°C.

(*b*) The quantity of heat stored in: (i) the concrete, and (ii) the fibreboard.

Data:

Thermal conductivity of concrete	= 1.44 W/m°C
Thermal conductivity of fibreboard	= 0.046" " "
Density of concrete	= 2 400 kg/m^3
Density of fibreboard	= 288 kg/m^3
Mean specific heat capacity of concrete	= 1.0 kJ/kg°C
Mean specific heat capacity of fibreboard	= 1.8 kJ/kg°C
Thermal resistance of air space	= 0.18 m^2°C/W
Inside surface resistance	= 0.12 " " "
Outside surface resistance	= 0.045" " "
Inside air temperature	= 21.1°C
Outside air temperature	= 1.7°C
Datum temperature	= −1.1°C (NC)

Ans.: (*a*) 19 mm; (*b*) (i) 1 828 kJ/m^2, (ii) 146.5 kJ/m^2 approx.

5. A building has a symmetrical pitched roof with vertical gable ends, the ridge being 3.66 m above a horizontal ceiling. The ceiling measures 12.2 m × 12.2 m and has a thermal transmittance (U) of 2.27 W/m^2°C. The thermal transmittance of the roof is 3.97 W/m^2°C and that of the gable ends 2.5 W/m^2°C.

If the attic space has four air changes per hour, and the air temperature beneath the ceiling is maintained at 15.6°C when outside air is at −1.1°C, calculate:

(*a*) Equilibrium air temperature in attic space.

(*b*) The heat loss through the ceiling.   (NC)

Ans.: (*a*) 3.3°C approx.; (*b*) 5.1 kW approx.

6. A 50 mm vertical air gap has an average temperature of 10°C and a temperature difference between the two faces of 5.6°C. If E_1 is the emissivity of one face, E_2 that of the other face, calculate the thermal resistance of the radiation component of heat transfer when:
(i) $E_1 = E_2 = 0.9$. By what ratio is this resistance increased if:
(ii) $E_1 = 0.1$ and $E_2 = 0.9$; (iii) $E_1 = E_2 = 0.1$?

If the thermal resistance of the convection component for the same gap is 0.81 m²°C/W, find the total resistance in each of the three cases. Hence draw conclusions relating to the use of bright aluminium foil.   (IHVE)*

Ans.: (i) 0.23 m²°C/W increased in the ratios, (ii) 8.3 and (iii) 15.5. Total resistance in each case: (i) 0.18, (ii) 0.57 and (iii) 0.67 m²°C/W.

Note. Use parallel-planes equation for radiation.

7. A structural flue from an oil-fired boiler installation is so sited that one wall is integral with the boundary of an occupied room. If the room air temperature is 18.3°C and the flue-gas temperature is 260°C, calculate the wall surface temperature in the room when the flue construction is, alternatively:

(*a*) Solid 460 mm brickwork with 16 mm plaster on the room surface.
(*b*) Solid 340 mm brickwork with a 25 mm air gap, 114 mm fibrebrick lining and 16 mm plaster on the room surface.
(*c*) Solid 340 mm brickwork with 114 mm molar brick lining bonded in and 16 mm-thick plaster on the room surface.
(*d*) Solid 460 mm brickwork faced on the room side with 50 mm studding to carry 13 mm-thick plasterboard over a cavity packed with glass fibre.

The following resistances and resistivities may be assumed

Resistances:

Hot surface film	= 0.09 m²°C/W
Room surface film	= 0.12 ” ” ”
Air gap	= 0.18 ” ” ”

Resistivities:

Common brick	= 0.87 m°C/W
Firebrick	= 0.93 ” ” ”
Molar brick	= 6.9 ” ” ”
Plaster	= 1.7 ” ” ”
Plasterboard	= 6.9 ” ” ”
Glass fibre	= 20.8 ” ” ”

Ans.: (*a*) 65°C; (*b*) 55°C; (*c*) 41°C and (*d*) 35°C approx.

* Refers throughout to data from the Institution of Heating and Ventilating Engineers.

8. (*a*) Draw to scale the temperature gradient across a 230 mm brick wall with 25 mm slab cork fixed to the internal surface. Conductivity or λ value to be taken as 1.15 and 0.05 W/m°C for brick and cork, and internal and external resistances Rs_1 and Rs_2 as 0.12 and 0.05 m²°C/W. Internal and external temperatures are 18.3°C and −1.1°C.

(*b*) The total heat losses in an office building have been calculated as 234.4 kW when the internal and external temperatures are 15.6°C and −1.1°C, Window areas in the building comprise 1 115 m² of single glazing with U value of 5.68 W/m²°C.

Calculate the maximum temperature attainable in the building if 60 per cent of the glass area is replaced with double glazing. Take 2.84 for U value for double glazing and assume heat input to the building and rate of air change remain constant.   (IHVE)

Ans.: approx.

    (*a*) Internal surface temperature   = 15.6°C
        Interface temperature   = 5°C
        External surface temperature   = 0°C
    (*b*) 17.8°C

9. A room has a width of 6.1 m and has a roof having a 45° pitch covered with tiles on battens. There is a 19 mm-thick horizontal plastered ceiling below.

Calculate the overall U value if 19 mm of insulation of λ = 0.04 is laid above the plaster.

Take: R_{S_1} = 0.096 and R_{S_2} = 0.044, λ for tiles = 1.15 and λ for plaster = 4.4.

Assume tiles 19 mm thick.   (IHVE)

Ans.: U = 1.25

10. A lift shaft passing through the open is formed by a steel tube 1.1 m internal diameter. Externally the shaft is covered with 0.15 m-thick cell concrete slabs and finished by a layer of 0.015 m-thick hard-setting concrete. Neglecting the thickness and thermal resistance of the steel tube, determine how much heat will have to be provided per cubic foot of shaft at an external air temperature of −1.1°C if an internal air temperature of 12.8°C with 1½ air changes per hour is to be maintained inside. Ends of shaft to be ignored.

    Internal surface resistance   = 0.12 m²°C/W
    External surface resistance   = 0.07 ” ” ”
    Thermal conductivity of concrete   = 1.44 W/m°C
    Resistivity of cell concrete   = 9.22 m°C/W

Ans.: 160 kJ/m³ approx.   (NC)

2: Heat Emission

Pipes, Radiators, Convectors, Unit Heaters, Embedded and Radiant Panels

Pipes

Although pipes fixed at skirting level provide uniform conditions of warmth, they are generally considered to be unsightly and are therefore rarely used as heating surfaces. Since pipes are, however, used for distributing hot water and steam, it is necessary to determine their heat emission, which may represent a significant fraction of the heat required by the units served.

The heat flow area is proportional to the radius, and therefore the temperature gradient is inversely proportional to the radius. Consider the radial flow of heat through a thin cylindrical element of thickness dr for a temperature difference of dt as shown in Fig. 2.1, then expressing the rate of heat flow (Φ), in terms of unit length of cylinder:

$$\frac{\Phi}{l} = -\lambda \, 2\pi r \frac{dt}{dr}$$

from which

$$\frac{\Phi}{2\pi\lambda l} \cdot \frac{dr}{r} = -dt$$

since $\dfrac{\Phi}{l}$ is the same for all radii

$$\frac{\Phi}{2\pi\lambda l} \int_{r_1}^{r_2} \frac{dr}{r} = -\int_{t_1}^{t_2} dt$$

and

$$\frac{\Phi}{2\pi\lambda l} \cdot \log_e \frac{r_2}{r_1} = [t_1 - t_2]$$

whence

$$\frac{\Phi}{l} = \frac{2\pi\lambda(t_1 - t_2)}{\log_e \dfrac{r_2}{r_1}} \qquad (2.1)$$

i.e.

$$\frac{\Phi}{l} = \frac{\Delta t}{R}$$

and

$$R = \frac{\log_e \dfrac{r_2}{r_1}}{2\pi\lambda}$$

If, alternatively, an arithmetic mean radius is used the approximate rate of heat flow per unit length of cylinder will be:

$$\frac{\Phi}{l} = \frac{\pi\lambda(r_1 + r_2)(t_1 - t_2)}{r_2 - r_1}$$ (2.2)

The exact rate of heat flow would then be:

$$\frac{\Phi}{l} = \frac{f.\lambda\pi(r_1 + r_2)(t_1 - t_2)}{r_2 - r_1}$$ (2.3)

where f = a correction factor.

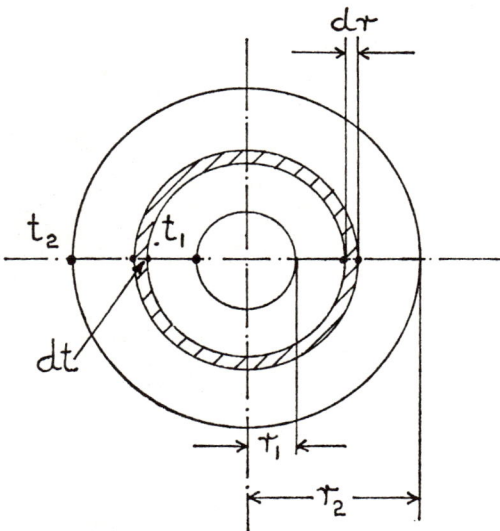

Fig. 2.1. Radial heat flow.

It may be shown that if the ratio r_2/r_1 is less than 2 the correction factor will have a value between 0.96 and 1.0.

Consider also the rate of heat transfer at the inside and outside surfaces for for unit length of cylinder.

Let h_{si} = conductance per unit area at inside surface,
h_{so} = conductance per unit area at outside surface,
t_i = mean temperature of fluid flowing inside pipe,
t_{S_i} = inside surface temperature of pipe,
t_{S_o} = outside surface temperature of pipe,
t_a = ambient air temperature of pipe,

Then, for steady heat flow conditions

$$\frac{\Phi}{l} = h_{si}\, 2\pi_{r_1}(t_i - t_{S_i})$$

also

$$\frac{\Phi}{l} = \frac{2\pi\lambda(t_{S_i} - t_{S_o})}{\log_e \frac{r_2}{r_1}}$$

also

$$\frac{\Phi}{l} = h_{so}\, 2\pi r_2(t_{S_o} - t_a)$$

from which

$$t_i - t_{S_i} = \frac{\Phi}{h_{si}2\pi r_1 l}$$

$$t_{S_i} - t_{S_o} = \frac{\Phi \log_e \frac{r_2}{r_1}}{2\pi\lambda l}$$

and

$$t_{S_o} - t_a = \frac{\Phi}{h_{so}2\pi r_2 l}$$

Then by addition

$$t_i - t_o = \frac{\Phi}{l}\left[\frac{1}{h_{si}2\pi r_1} + \frac{\log_e \frac{r_2}{r_1}}{2\pi\lambda} + \frac{1}{h_{so}2\pi r_2}\right]$$

and

$$\frac{\Phi}{l} = \frac{t_i - t_o}{\dfrac{1}{h_{si}2\pi r_1} + \dfrac{\log_e \frac{r_2}{r_1}}{2\pi\lambda} + \dfrac{1}{h_{so}2\pi r_2}} \qquad (2.4)$$

For a pipe with composite insulation of *n* layers Eq (2.4) becomes

$$\frac{\Phi}{l} = \frac{t_i - t_o}{\dfrac{1}{h_{si}2\pi r_1} + \dfrac{\log_e \frac{r_2}{r_1}}{2\pi\lambda_1} + \dfrac{\log_e \frac{r_3}{r_2}}{2\pi\lambda_2} + \cdots \dfrac{\log_e \frac{r_n}{r_{n-1}}}{2\pi\lambda_n} + \dfrac{1}{h_{so}2\pi r_n}} \qquad (2.5)$$

Note. λ and r must be in consistent units.

Since the thermal conductivity of the pipe wall is relatively high and the thickness small, the resistance will be very small and may be neglected when applying Eq (2.5).

The heat emission from the external surface takes place by radiation to the cooler surrounding surfaces and by convection to the ambient air, and therefore depends upon several factors, such as nature, orientation and temperature of the surface, air movement and temperature difference. Since the combined surface conductance, h_{so}, in Eq (2.5) is used in conjunction with the ambient air temperature t_a,

$$h_{so} = \frac{\Phi_{r/A} + \Phi_{c/A}}{t_{s_o} - t_a}$$

where

Φ_r and Φ_c = heat emission by radiation and convection respectively, and A = area.

It is usual to determine h_{so} for freely exposed horizontal pipes in still air and, for convenience, to assume that the surrounding air and mean surface temperatures are the same. Values of h_{so} covering a wide range of temperature differences and pipe diameters are given in the *Guide to Current Practice, 1970*. Correction factors are also given for vertical pipes, pipes against walls or ceilings and for banks of pipes. The heat emission from insulated pipes is given as a percentage of the bare pipes loss.

Radiators

Modern radiators, available in a wide variety of heights and widths, are made from cast iron or pressed steel. The heating effect obtained is due largely to free convection currents set up within the room, and for this reason radiators should not be installed in recesses where air circulation would be restricted. The comparatively small amount of heat emitted by relatively low temperature radiation is not significantly affected by painting radiators with ordinary paints and enamels regardless of their colour. Metallic paints, however, may reduce the total heat emission by as much as 15 per cent. For radiators in recesses the total heat emission may be reduced by 10–20 per cent, depending on whether the recess is open or closed with a grille at the front. Wherever possible, radiators should be in-installed beneath windows, for with this arrangement objectionable cold down draughts may be eliminated. Where radiators are placed against internal partitions and walls it is advisable to fit a shelf not less than 100 mm above them to prevent discoloration of the wall decoration; the total heat emission is then reduced by about 4 per cent. The main disadvantage with radiator heating systems is the rather distinct air temperature gradient set up in the room from floor to ceiling.

The rated output of radiators given in manufacturers' catalogues are generally determined from the minimum tests specified in B.S.3528:1962. The total heat emission from radiators is generally considered to follow the empirical law:

$$\frac{\Phi_1}{\Phi_2} = \left(\frac{\Delta t_1}{\Delta t_2}\right)^{1.3} \tag{2.6}$$

where

Φ = heat emission,
Δ*t* = difference between the ambient air temperature and the mean surface temperature for the radiator.

For hot-water systems the mean surface temperature is taken as the arithmetic mean of the flow and return water temperatures.

Convectors

These appliances operate by natural convection, and consist essentially of a finned or gilled tube heater element encased at low level in a sheet-steel cabinet provided with inlet and outlet air grilles at the bottom and top respectively. The heater element may be arranged for use with either low-pressure hot water, high-pressure hot water or steam. With convectors the heating effect is due practically entirely to air movement, the thermal radiation component being very small indeed. Convectors are suitable for fixing in wall recesses, or may be installed free standing beneath windows or on the face of walls. This form of heating creates a considerable temperature gradient within the room. The total heat emission from convectors depends upon several factors, such as height of air outlet above the floor, whether the outlet grille is at the top of the cabinet or on the front face, and upon the temperature difference between the entering air and the heater element. The rated output may be determined from the minimum tests specified in B.S.3528 : 1962. The total heat emission from convectors is generally considered to follow the empirical law.

$$\frac{\Phi_1}{\Phi_2} = \left(\frac{\Delta t_1}{\Delta t_2}\right)^{1.5} \tag{2.7}$$

where

Φ = heat emission,
Δ*t* = difference between entering room air and mean surface temperatures.

Comparing this with Eq (2.6), it will be seen that the heat emission from convectors decreases more rapidly than for radiators. For this reason convectors should be served and controlled by a separate circuit, and not connected in with radiators.

Unit Heaters

These consist essentially of an electrically driven fan unit that blows cool room air over a heater battery. The warmed air is then discharged via a set of louvres into the room. Unit heaters for mounting at high level are arranged to discharge the warm air either diagonally downwards or vertically downwards; these are particularly suitable for relatively high rooms. A fairly high discharge velocity is essential to force the warm air downwards against its natural tendency to rise. Units are available in a variety of types and range of output. Very large output units are usually of the floor-standing type. Unit heaters are particularly suitable for heating factory buildings, where they may be suspended from the roof trusses. In summer when the heat is not required the fan unit may be run to circulate the room air. Unit heaters can also be obtained with fresh-air inlets or with part fresh- and part room-air inlets. They are available with either hot-water, steam, gas or electric heater battery elements.

The heat output of unit heaters depends upon many factors, such as: rate of air flow, temperature of heating medium, type of heating surface and type of finning, number and arrangement of the rows of tubes and upon the temperature of the entering air. Rated outputs covering a wide variation in the above factors are generally given in manufacturers' lists, which also suggest suitable mounting heights and spacings.

In general the heat output of unit heaters will be:

$\Phi = U_i A_i \Delta t_m$ for steam heated elements and,

$\Phi = U_i A_i f(\Delta t_m)$ for hot water heated elements where:

U_i = overall heat transfer coefficient of the element related to inside area (A_i) of the tubes,

Δt_m = logarithmic mean temperature difference between the heating media and the air flowing over the element,

f = a correction factor to allow for the relative direction of the water and air flow.

It may be shown that the overall heat transfer coefficient (U_i) is given by:

$$U_i = \left[\frac{1}{h_{si}} + \frac{1}{h_{so}\left(\frac{A_F}{A_i}\phi + \frac{A_b}{A_i}\right)} \right]^{-1}$$

where h_{si} and h_{so} = inside and outside surface conductances respectively,

A_F = fin area

A_b = bare tube area between fins

A_i = inside tube area

ϕ = fin efficiency.

With steam heated unit heaters the value of h_{si} is comparatively large and may be neglected. U_i may therefore be considered to be proportional to h_{so} which may be determined from: $h_{so} = C R_e^{0.6} P_r^{0.3}$ which is similar to Eq Int.15 and in this case should be based on the minimum air velocity through the tube arrangement. The value of C depends on several factors such as the number of rows and on their arrangement, i.e. staggered or in-line, and also on the Reynolds number; a typical value is 0.3. For a particular heater element $h_{so} \propto R_e^{0.6}$.

Consider an in-line tube arrangement and let:

u_1 = face velocity of air
u_2 = velocity of air between rows
A_1 = face area of heater element
A_2 = area of space between two adjacent tubes
n = number of spaces in a row
ρ = density of air.

Then

$$A_1 \, u_1 \, \rho = n \, A_2 \, u_2 \, \rho$$

Since A_1, A_2 and ρ are constant, u_1 is proportional to u_2 and h_{so} will be proportional to $u_1^{0.6}$, i.e.

$$U_i \propto u_1^{0.6}$$

Since the air volume entering the heater element is proportional to the fan speed (N) then: $u_1 \propto N$ and therefore:

$$U_i \propto N^{0.6}$$

A similar solution would be obtained by considering a staggered tube arrangement.

Let t_1 = entering air temperature
t_2 = leaving air temperature
t_s = steam temperature
Δt_m = logarithmic mean temperature difference
Φ = heat output of heater element

Then for a constant overall thermal transmittance coefficient $\Phi \propto \Delta t_m$ and for a constant rate of air flow and specific heat capacity $\Phi \propto (t_2 - t_1)$.

Since

$$\Delta t_m = \frac{(t_s - t_1) - (t_s - t_2)}{\log_e \dfrac{t_s - t_1}{t_s - t_2}}$$

then

$$\Phi \propto \frac{t_2 - t_1}{\log_e \dfrac{t_s - t_1}{t_s - t_2}}$$

also

$$(t_s - t_2) \propto (t_s - t_1)$$

In the case of hot water heated elements h_{si} is not too large to be neglected and may be determined using Eq Int.16. With hot water elements no simple relation exists between U_1 and u_1 and U_1 should be determined as shown earlier.

Embedded Pipe Panel Heating

With this method of heating, water at a comparatively low mean temperature is circulated through pipe coils embedded beneath the surface of ceilings or floors and sometimes walls. The boundary surfaces of the room are thus the heat-emitting surfaces, and the effectiveness of the system therefore depends to a great extent on obtaining a correct combination of air temperature, panel temperature and mean temperature of the unheated surfaces of the room.

Current interest in floor heating is undoubtedly due to the increasing use of embedded electric heating cables, often using off-peak electricity, as a competitor to embedded hot-water pipes. While floor heating has been used for about fifty years, there is still a dearth of information on the effect on the floor emission of pipe spacing and diameter, depth of cover, floor finish and intermittent operation. The mean surface temperature of floor panels is generally limited to 25°C maximum for continuously occupied rooms and 30°C for corridors, foyers, banking halls and similar spaces. The use of temperatures above these maximum values will result in considerable discomfort to the feet. As a result of this restriction on surface temperature, the heat emission from floor panels may not in some cases be sufficient to meet the heat requirements of the room, and supplementary heating may be required. This is particularly the case with single-storey buildings, but clearly depends on the amount of glazing and degree of structural insulation. Floor coils are used mostly in solid ground floor, the downward and edge loss depending on the plan aspect ratio of the building and the thermal conductivity of the subsoil. The downward and edge loss, which may be as much as 30 per cent of the total heat input, may be reduced by using thermal insulation beneath the panel coils. A 1 m-wide horizontal strip of insulation placed around the perimeter of the floor slab is all that is necessary in most cases, but with large plan aspect ratios, above say 4.0, the use of insulation over the entire floor area should be considered.

In multi-storey buildings intermediate floors are heated by the combined upward emission from the floor and the downward emission, through the ceiling, from the floor panel above. The proportions of upward and downward emission depends on the respective thermal resistances of the two heat-flow paths. When comparing one type of construction with another it is sufficient to consider a mean temperature (t_p) at the level of the embedded heating pipes (or cables). Taking a simple case of a 150 mm solid concrete floor with 50 mm surface screed and assuming 18°C air temperature above and below, a floor surface temperature of 25°C and an upward emission of 9 W/m^2°C it may be shown that 64 per cent of the total heat input is upwards and 36 per cent downwards, the ceiling temperature being about 23°C. The downward emission may be reduced by the use of thermal insulation placed just below the pipes or cables. Assuming that the insulation has a resistance of 0.53 m^2°C/W, the proportions become 84 per cent upwards and 16 per cent downwards, giving a ceiling temperature of about 20°C. The laying of underfelts and carpets on intermediate floors will reduce the upward emission and increase the downward emission. The interaction between floors of multi-storey buildings will be reduced considerably, if insulation is used below the pipes or cables.

The permissible surface temperature of ceiling panels is limited by the maximum intensity of radiation which may be allowed on the heads of the room occupants, and therefore on the height and plan aspect ratio of the room. An occupant beneath the centre of the panel would receive the maximum amount of radiation. Ceiling panels are normally able to meet the heat requirements of the room when operating at surface temperatures lower than the maximum permissible for comfort. While in most cases ceiling panels cover the entire ceiling area of normal rooms, they are sometimes used as strip panels above tall windows to counteract excessive radiation from the body. In multi-storey building there will be some upward emission which should be taken into account; see notes given above for floor panels in multi-storey buildings.

Embedded pipe and cable panels may be used effectively and economically in all types of buildings of traditional and non-traditional construction, provided that interaction between floors is kept to a minimum. With ceiling panels the air temperature is practically uniform between the floor and ceiling. Floor and wall panels give only a slight temperature gradient. An allowance for height is therefore unnecessary when calculating room heat requirements. Owing to the comparatively high radiation component, comfort conditions are obtained within the room with an air temperature a few degrees lower than with convective heating methods. Since the panels and attendant pipework are completely embedded in the building construction, there are no obstructions in the rooms and partitions, and furniture may be placed exactly where required. The capital cost is usually higher than for other heating methods, but the fuel costs are usually less.

The heat emission from embedded pipes and cables is determined generally in accordance with the data given in "Introduction to Elementary Heat Transfer" and as shown in the examples that follow later in this chapter.

Radiant Panels

These panels consist essentially of a sinuous type of steel pipe coil welded to a standard 2 m X 1 m or 2.4 m X 1.2 m heavy sheet steel plate which receives heat from the coil by conduction. Heat is emitted by radiation and convection from both sides of the panel; i.e. directly from the pipe coil and indirectly from the plate. Radiant panels are also available with steel plates welded to both sides of the pipe coil. With this type the plates are heated mostly by conduction and partly by radiation from the coil sandwiches between them. Radiant panels are designed to use steam or high-pressure hot water as the heating medium, and normally operate at fairly high surface temperatures. For this reason they are not suitable for use in rooms which have low ceilings. They are particularly suitable for use in industrial buildings and are frequently called "industrial radiant panels".

Radiant panels are normally installed vertically and freely exposed in lines running the length of the building so that full advantage may be taken of the emission which takes place from both sides of the panel. They may, however, be installed vertically against outside walls or horizontally at high level in tall buildings, but in both these cases it is necessary to insulate the back and upper surface of the panels respectively to reduce unnecessary heat loss.

Another form of radiant panel consists essentially of a narrow strip of metal attached to a single 32 mm or 40 mm diameter steel pipe, made up in lengths of 7 m. These units are joined together to form a continuous-strip radiant heater. They are suitable for fixing vertically or horizontally. Multi-pipe types are also available.

With all radiant panel-heating methods a considerable proportion of the heat emitted is radiant heat. This results in many cases in negligible air movement, and hence temperature gradient and possibly some fuel saving. Apart from their general use for the complete heating of factory buildings, radiant panels may be used in open and exposed positions to provide spot or local heating.

Example 2.1. A pipe of 100 mm outside diameter is insulated with a layer of 50 mm-thick plastic insulation for which the thermal conductivity is 0.05 W/m°C. The pipe conveys steam at 180°C and runs through a room at 20°C. Taking the conductance at the inside of the pipe and the outside of the insulation as 850 and 12 W/m²°C respectively, calculate:

(*a*) the rate of heat flow, W/m;
(*b*) the temperature at the outer surface of the insulation;
(*c*) the temperature half-way through the insulation;
(*d*) the radius at which the temperature is 100°C.

Neglect the thickness and resistance of *the pipe wall.*

(*a*) Referring to Eq (2.5), in this example:

r_1 = 0.05 m
r_2 = 0.1 m
r_3 (radius to mid-point of insulation) = 0.075 m

Resistance at inside surface

$$= \frac{1}{850 \times 2\pi \times 0.05} = 0.0037 \text{ m°C/W}$$

Resistance at outside surface

$$= \frac{1}{12 \times 2\pi \times 0.1} = 0.133 \quad \text{''}$$

Resistance of insulation

$$= \frac{\log_e \dfrac{0.1}{0.05}}{2\pi \times 0.05} \qquad = 2.206 \qquad \text{''}$$

Resistance to mid-point of insulation

$$= 0.0037 + \frac{\log_e \dfrac{0.075}{0.05}}{2\pi \times 0.05} = 1.29 \qquad \text{''}$$

Note relatively small value of inside surface resistance, which in many cases may be neglected

Then

$$\frac{\Phi}{l} = \frac{180 - 20}{0.0037 + 2.206 + 0.133} = 68.28 \text{ W/m}$$

(*b*) Since $\dfrac{\Phi}{l} = \dfrac{\Delta t}{R}$

then $\qquad \Delta t = \dfrac{\Phi}{l}.\,R$

and $\qquad t_{S_o} = 20 + 68.28 \times 0.133 = 29°C$

(*c*) $\qquad t_{\text{mid}} = 180 - 68.28 \times 1.29 = 92°C$

(*d*) Let $\quad R_{100}$ = resistance at radius r_{100}, where $t = 100°C$ and

$\qquad R_x$ = resistance of insulation at same point

Then

$$180 - 68.28\,.\,R_{100} = 100$$

from which

$$R_{100} = 1.171 \text{ m}°C/W$$

Since $\qquad R_{100} = R_{S_i} + R_x$

$$R_x = 1.171 - 0.0037 = 1.167 \text{ m}°C/W$$

that is

$$\frac{\log_e \dfrac{r_{100}}{0.05}}{2\pi \times 0.05} = 1.167$$

$$\log_e \frac{r_{100}}{0.05} = 0.3667$$

from which $\qquad\qquad r_{100} = 0.072 \text{ m}$

$$= 72 \text{ mm}$$

Example 2.2. A pipe of 115 mm outside diameter conveys high-pressure hot water at a mean temperature of 150°C. Calculate the thickness of insulation that should be applied to the pipe to prevent the outside surface exceeding a temperature of 40°C for a heat loss of 96 W/m. Take the thermal conductivity of insulation at 0.07 W/m°C.

Applying Eq (2.1),

$$96 = 2\pi \times 0.07 \frac{(150 - 40)}{\log_e \frac{2r_2}{115}}$$

$$\log_e \frac{2r_2}{115} = 2\pi \times 0.07 \times \frac{110}{96}$$

$$= 0.504$$

$$\frac{2r_2}{115} = 1.655$$

$$r_2 = 95.2 \text{ mm}$$

$$\therefore \text{ thickness of insulation} = 95.2 - \frac{115}{2}$$

$$= 38 \text{ mm} \qquad \text{Ans.}$$

Example 2.3. During a test the external surface temperature of an insulated pipe was found to be 32°C when exposed to air at 24°C and surrounding surfaces having a mean temperature of 18°C. Calculate the rate of heat loss from the pipe if the outside diameter of the insulation is 200 mm.

In this case

$$\frac{\Phi}{A} = \frac{\Phi_r}{A} + \frac{\Phi_c}{A}$$

where $\dfrac{\Phi_r}{A}$ = heat emission by radiation

$\dfrac{\Phi_c}{A}$ = heat emission by convection.

For most cases, see Eq (Int.35).

$$\frac{\Phi_r}{A} = 5.67 \, F_A \cdot F_E \left[\left(\frac{T_1}{100} \right)^4 - \left(\frac{T_2}{100} \right)^4 \right]$$

or

$$\frac{\Phi_r}{A} = 5.67 \, E \left[\left(\frac{T_1}{100} \right)^4 - \left(\frac{T_2}{100} \right)^4 \right]$$

where $E = F_A \cdot F_E$ which may be taken as 0.9
T_1 and T_2 = absolute temperature of external surface and surrounding
      surfaces respectively.

Then $\dfrac{\Phi_r}{A} = 5.67 \times 0.9 \left[\left(\dfrac{273 + 32}{100} \right)^4 - \left(\dfrac{273 + 18}{100} \right)^4 \right]$

$= 75.7 \text{ W/m}^2$

The following approximate expression may be used for calculating the
heat emission by convection:

$$\frac{\Phi_c}{A} = \frac{1.32(t_{so} - t_a)^{1.25}}{D_o^{0.25}}$$

where t_{s_o} = external surface temperature, °C
      t_a = ambient air temperature, °C
      D = outside diameter, m

Then $\dfrac{\Phi_c}{A} = \dfrac{1.32(32 - 24)^{1.25}}{0.2^{0.25}}$

$= 26.6 \text{ W/m}^2$

Total heat emission

$= 75.7 + 26.6 = 102.3 \text{ W/m}^2$

or alternatively

$= 102.3 \times 0.8\pi = 257 \text{ W/m}$

Example 2.4. What is meant by the economic thickness of pipe insulation?

While for any specific purpose the thermal conductivity of the material
governs the insulation thickness, the money spent on the insulation should
be weighed against the saving in fuel. The annual cost of insulation depends
on its cost and depreciation rate, and increases with increase in thickness.
On the other hand, the annual cost of the heat lost depends on the heat loss
per hour, the hours of use per annum and on the basic cost of heat, and
decreases with increase in thickness. Therefore the thickness of insulation
which results in the minimum total annual cost will be the economic thickness. Two methods of determining economic thickness are given in
B.S.1588:1949; they are the minimum-cost method and the incremental-
payment method. The basic principle is shown in Fig. 2.2.

Example 2.5. Fig. 2.3 shows hot water radiators connected: (a) to a one-
pipe system; (b) to a two-pipe system; and (c) in series. Using the data
listed below, determine the total number of radiator sections required for
each circuit arrangement.

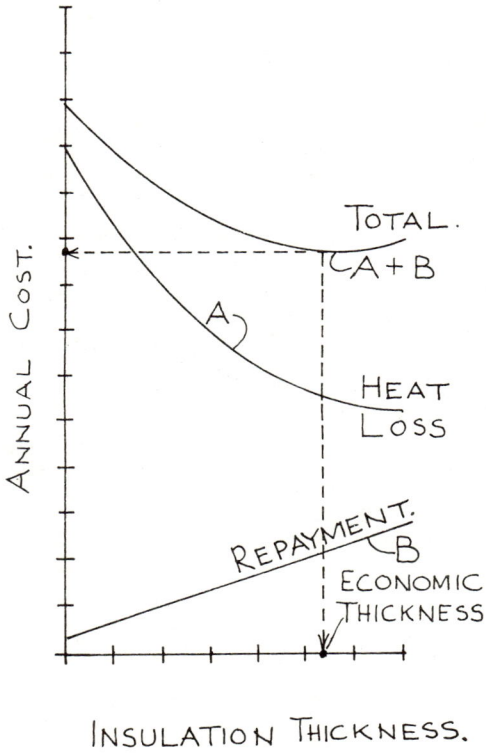

Fig. 2.2 Economic thickness of insulation.

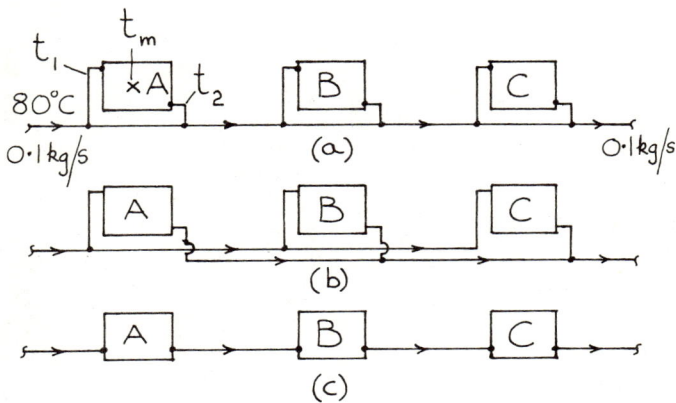

Fig. 2.3. Radiator connections.

Data:—

Initial water temperature in each case	$= 80°C$
Final " " " " "	$= 60°C$
Room temperature	$= 20°C$
Emission required from each radiator	$= 2.8 \text{ kW}$
Rated output of the radiators	$= 100 \text{ W per}$

section for $55°C$ difference between the room and mean radiator temperatures

Temperature drop across each radiator in case (a)	$= 10°C$
Specific heat capacity of water	$= 4.2 \text{ kJ/kg}°C$

Let t_a = room air temperature, $°C$
  t_1 = water temperature at inlet to radiator, $°C$
  t_2 = water temperature at outlet from, $°C$
  t_m = mean surface temperature of radiator = $\frac{1}{2}(t_1 + t_2)$, $°C$
  Δt = temperature difference, $t_m - t_a$, $°C$
  N = number of radiator sections taken to the next whole number
  Φ = heat output W per section

Consider arrangement (*a*).
Water flow rate at beginning of circuit will be:

$$\frac{3 \times 2.8}{4.2(80 - 60)} = 0.1 \text{ kg/s}$$

Water flow rate per radiator will be:

$$\frac{2.8}{4.2 \times 10} = 0.067 \text{ kg/s}$$

Radiator A:

$$t_1 = 80°C$$
$$t_2 = 80 - 10 = 70°C$$
$$t_m = \tfrac{1}{2}(t_1 + t_2) = 75°C$$
$$\Delta t = 75 - 20 = 55°C$$

for which $\Phi = 100 \text{ W/section}$

and $N = \dfrac{2.8 \times 10^3}{100} = 28 \text{ sections}$

Radiator B:

Assuming that the specific heat capacity of the water is constant then:

$$t_1 = \frac{(0.067 \times 70) + 80(0.1 - 0.067)}{0.1} = 73.3°C$$

alternatively, by calculation of temperature drop:

$$t_1 = 80 - \frac{2.8}{0.1 \times 4.2} = 73.3°C \text{ as before}$$

then, $t_2 = 73.3 - 10 = 63.3°C$

and $t_m = 68.3°C$

alternatively, by calculation of temperature drop below t_m of radiator A:

$$t_m = 75 - \frac{2.8}{0.1 \times 4.2} = 68.3°C \text{ as before}$$

$$\Delta t = 68.3 - 20 = 48.3°C$$

Using Eq (2.6),

$$\Phi = 100\left(\frac{48.3}{55}\right)^{1.3} = 85 \text{ W/section approx.}$$

then $N = \dfrac{2.8 \times 10^3}{85} = 33 \text{ sections}$

Radiator C:

$$t_m = 68.3 - \frac{2.8}{0.1 \times 4.2} = 61.6°C$$

$$\Delta t = 61.6 - 20 = 41.6°C$$

$$\Phi = 100\left(\frac{41.6}{55}\right)^{1.3} = 70 \text{ W/section approx.}$$

then $N = \dfrac{2.8 \times 10^3}{70} = 40 \text{ sections}$

Total number of sections for arrangement (*a*)

$$= 28 + 33 + 40 = 101 \text{ sections}$$

Consider arrangement (*b*)
Since each radiator will receive water at 80°C and because the final temperature is 60°C, the mean surface temperature of each radiator will be 70°C and Δt will be $70 - 20 = 50°C$ for which: –

$$\Phi = 100\left(\frac{50}{55}\right)^{1.3} = 88.5 \text{ W/section approx.}$$

then $N = \dfrac{2.8 \times 10^3}{88.5} = 32 \text{ sections/radiator}$

Total number of sections for arrangement (*b*)

$$= 3 \times 32 = 96 \text{ sections}$$

Consider arrangement (*c*).
Since the water flow rate is 0.1 kg/s, see (*a*) above, the temperature drop

across each radiator will be $\dfrac{2.8}{0.1 \times 4.2}$ = 6.7°C and the mean surface tempera-

ture of the radiators will be:

$$\text{Radiator A} = 80 - \frac{6.7}{2} = 76.7°\text{C}$$

$$\text{''}\quad \text{B} = 76.7 - 6.7 = 70°\text{C}$$

$$\text{''}\quad \text{C} = 70 - 6.7 = 63.3°\text{C}$$

then

$$\Delta t_A = 76.7 - 20 = 56.7°\text{C}$$

$$\Delta t_B = 70 - 20 = 50°\text{C}$$

$$\Delta t_C = 63.3 - 20 = 43.3°\text{C}$$

and

$$\Phi_A = 100 \left(\frac{56.7}{55}\right)^{1.3} = 104 \text{ W/section, approx.}$$

$$\Phi_B = 100 \left(\frac{50}{55}\right)^{1.3} = 88.5 \text{ ''}\quad\text{''}\quad\text{''}$$

$$\Phi_C = 100 \left(\frac{43.3}{55}\right)^{1.3} = 73.2 \text{ ''}\quad\text{''}\quad\text{''}$$

then

$$N_A = \frac{2.8 \times 10^3}{104} = 27 \text{ sections}$$

$$N_B = \frac{2.8 \times 10^3}{88.5} = 32 \text{ ''}$$

$$N_C = \frac{2.8 \times 10^3}{73.2} = 39 \text{ ''}$$

Total number of sections for arrangement (*c*)

$$= 27 + 32 + 39 = 98 \text{ sections.}$$

Summary:

    (*a*) one-pipe arrangement: 101 sections
    (*b*) two-pipe arrangement: 96 sections
    (*c*) series arrangement: 98 sections

Example 2.6. A hot-water radiator heating system is designed to operate with flow and return water temperatures of 82°C and 66°C respectively for inside and outside air temperatures of 18°C and −1°C respectively. The heating system is supplied with hot water from a boiler plant working, and supplying other loads, at a constant temperature of 82°C, a three-way valve being included in the system. Analyse the water temperatures and flow rates for the heating system for outside air.temperatures of $-1°C$, $4°C$ and $7°C$ if the three-way valve is used:

(a) as a flow-mixing system of control;
(b) as a flow-diverting system of control.

Let t_a = inside air temperature
 t_o = outside air temperature
 t_1 = boiler water temperature
 t_2 = system flow water temperature
 t_3 = system return water temperature
 t_m = mean surface temperature of radiators = $\frac{1}{2}(t_2 + t_3)$
 $\dot{m}_1$ = water flow rate
 $\dot{m}_2$ = water flow rate in system by-pass
 x = water flow rate in boiler by-pass, as a fraction
 Φ = system heat output
 C = constant
 c = specific heat capacity of the water.

Assuming steady heat flow conditions:
Heat loss from building = Heat emission from radiators = Heat given up by the water.

i.e. $$\Phi = C(t_a - t_o) = C_1(t_m - t_a)^{1.3} = \dot{m}_1 c(t_2 - t_3) \qquad (2.8)$$

(a) *Flow Mixing System of Control.* The general arrangement of the circuit is shown in Fig. 2.4.
For $t_o = -1°C$ and applying Eq (2.8).

$$\Phi = C(18 + 1) = C_1(74 - 18)^{1.3} = \dot{m}_1 c(82 - 66) \qquad (2.9)$$

Since $t_2 = t_1$ there will be no flow in the by-pass, i.e. $x = 0$.
For $t_o = 4°C$ we have:

$$\Phi = C(18 - 4) = C_1(t_m - 18)^{1.3} = \dot{m}_1 c(t_2 - t_3) \qquad (2.10)$$

Then from Eq (2.9) and Eq (2.10),

$$\frac{19}{14} = \left(\frac{56}{t_m - 18}\right)^{1.3}$$

from which $t_m = 63°C$ approx.
Also from Eq (2.9) and Eq (2.10),

$$\frac{19}{14} = \frac{16}{t_2 - t_3}$$

from which $t_2 - t_3 = 12°C$ approx.
and since $t_m = \frac{1}{2}(t_2 + t_3)$; $t_2 = 63 + 6 = 69°C$
and $t_3 = 63 - 6 = 57°C$

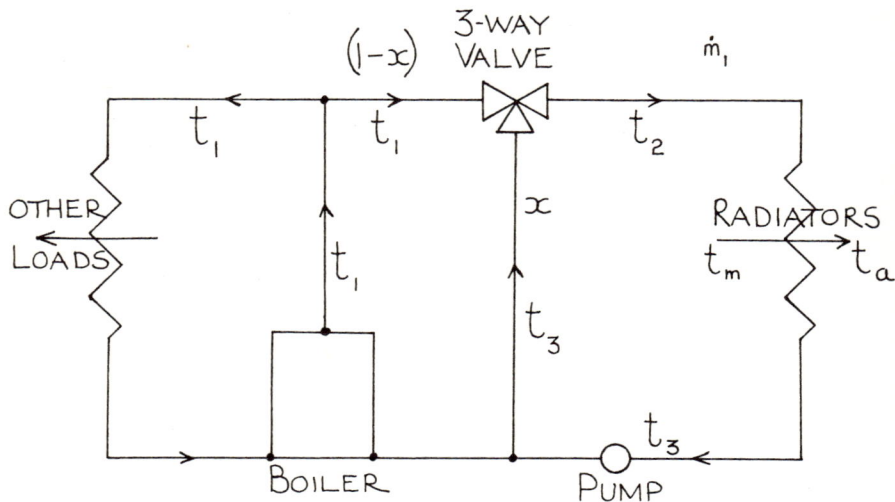

Fig. 2.4

Then by method of mixtures, and assuming that the specific heat capacity of the water is constant, we have

$$x \cdot 57 + (1 - x)82 = 69$$

from which $$x = 0.52$$

and $$1 - x = 0.48$$

For $t_o = 7°C$

$$\Phi = C(18 - 7) = C_1(t_m - 18)^{1.3} = \dot{m}_1 c(t_2 - t_3) \qquad (2.11)$$

Then from Eq (2.9) and Eq (2.11)

$$\frac{19}{11} = \left(\frac{56}{t_m - 18}\right)^{1.3}$$

from which $t_m = 55°C$.

Also from Eq (2.9) and Eq (2.11)

$$\frac{19}{11} = \frac{16}{t_2 - t_3}$$

from which $t_2 - t_3 = 9.3°C$

$$t_2 = 55 + 4.65; \text{ say} = 60°C$$

and $$t_3 = 55 - 4.65; \text{ say} = 50°C$$

Then by method of mixtures,

$$x \cdot 50 + (1 - x)82 = 60$$

from which $\qquad\qquad x = 0.69$

and $\qquad\qquad 1 - x = 0.31$

(b) *Flow Diverting System of Control.* The general arrangement of the circuit is shown in Fig. 2.5. In this case the water flow rate from the boiler is constant.

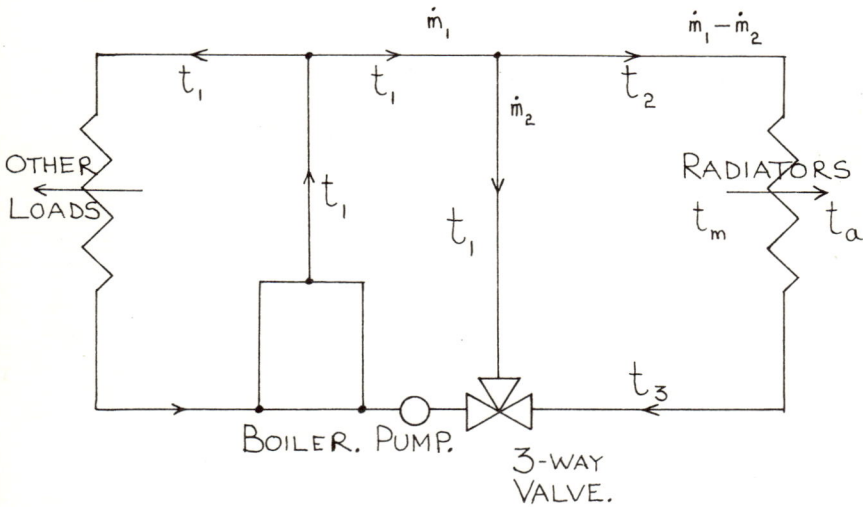

Fig. 2.5

For $\qquad t_o = -1°C, t_2 = t_1$ and $\dot{m}_2 = 0$

For $\qquad t_o = 4°C, t_m = 63°C$ (from part (a))

Since $\qquad t_2 = 82°C$ and $t_m = \frac{1}{2}(t_2 + t_3)$

then $\qquad t_3 = 44°C$

and $\qquad \dfrac{19}{14} = \dfrac{\dot{m}_1 (82 - 66)}{(\dot{m}_1 - \dot{m}_2)(82 - 44)}$

from which $\qquad \dfrac{\dot{m}_2}{\dot{m}_1} = 0.69$

For $\qquad t_o = 7°C, t_m = 55°C$ (from part (a))

Then since $t_2 = 82°C; \ t_3 = 28°C$

then $\dfrac{19}{11} = \dfrac{\dot{m}_1 (82 - 66)}{(\dot{m}_1 - \dot{m}_2)(82 - 28)}$

from which $\dfrac{\dot{m}_2}{\dot{m}_1} = 0.83$

Summary of results:—

(a) Flow mixing control:

t_0	t_1	t_2	t_3	x	$1 - x$
−1	82	82	66	0	1
4	82	69	57	0.52	0.48
7	82	60	50	0.69	0.31

(b) Flow diverting control:

t_0	t_1	t_2	t_3	$\dot{m}_2$ as per cent of $\dot{m}_1$
−1	82	82	66	0
4	82	82	44	69
7	82	82	28	83

Example 2.7. A hot-water radiator under test is controlled to operate at a constant mean temperature of 71°C and maintains a room at 16°C when 0°C outside. What inside temperature could be maintained for an outside temperature of 5°C?

Assuming steady heat flow:

Heat loss from room = Heat emission from radiator

If t_a = temperature of room

$$\frac{t_a - 5}{16} = \left(\frac{71 - t_a}{71 - 16}\right)^{1.3}$$

that is $$\frac{t_a - 5}{16'} = \left(\frac{71 - t_a}{55}\right)^{1.3}$$

Assuming three values for t_a:

t_a	$t_a - 5$	$71 - t_a$	$\dfrac{t_a - 5}{16}$	$\left(\dfrac{71 - t_a}{55}\right)^{1.3}$
18	13	53	0.81	0.95
20	15	51	0.94	0.91
22	17	49	1.06	0.86

Then by plotting the curves, Fig. (2.6),

    (*a*) Room heat loss $v . t_a$
    (*b*) Radiator emission $v . t_a$

the intersection gives the room temperature at which heat loss equals radiator emission and is found to be 19.5°C; the heat flow being 0.91 of the original value.

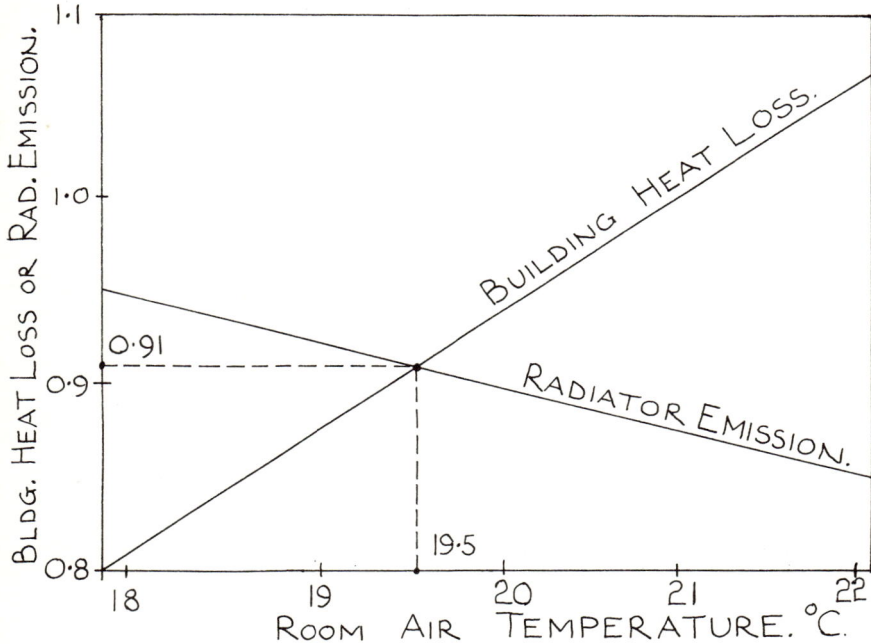

Fig. 2.6. Graphical solution of heat balance

Example 2.8. A building having a heat requirement of 150 kW is maintained at 16°C when 0°C outside by steam convectors having a mean temperature of 105°C. Subsequently it is required to maintain the building at 21°C when 0°C outside. Determine the mean temperature at which the convectors should operate to meet the increased heat requirement of the building.

Assuming steady heat flow.
Let t_m = mean temperature of convectors.

Then, since building heat requirement equals heat emission for convectors and using the 1.5 power law Eq (2.7),

$$\frac{150 \times 21}{16} = 150 \left(\frac{t_m - 21}{105 - 16} \right)^{1.5}$$

$$\left(\frac{t_m - 21}{89} \right)^{1.5} = \frac{21}{16}$$

From which $t_m = 128°C$

Example 2.9. A room is maintained at 18°C when −1°C outside by hot-water convectors working with inlet and outlet water temperatures of 94°C and 70°C respectively. How much additional convector heating surface should be installed to enable the room to be maintained at 24°C when−1°C outside with the same water temperatures.

Let A = total heating surface.

Then proceeding as above and using Eq (2.7)

$$t_m = \tfrac{1}{2}(94 + 70) = 82°C$$

$$\frac{18 + 1}{24 + 1} = \frac{A_1}{A_2} \left(\frac{82 - 18}{82 - 24} \right)^{1.5}$$

$$\frac{19}{25} = \frac{A_1}{A_2} \left(\frac{64}{58} \right)^{1.5}$$

from which $\dfrac{A_1}{A_2} = 0.655$ and $\dfrac{A_2}{A_1} = 1.53$

The additional heating surface required, expressed as a percentage of the original amount, will be

$$\frac{A_2 - A_1}{A_1} = \frac{A_2}{A_1} - 1$$

$$= 1.53 - 1 = 0.53 \text{ or } 53 \text{ per cent}$$

Example 2.10. A building is to be maintained at 18°C when −1°C outside by a hot-water heating system operating with flow and return water temperatures of 82°C and 60°C respectively. Calculate and compare the flow and return water temperatures required to maintain 18°C inside when 7°C outside if the heating appliances are: (*a*) radiators, and (*b*) convectors.

Applying Eq (2.15)

(*a*) Radiators, for $t_m = \tfrac{1}{2}(82 + 60) = 71°C$

$$\frac{18 + 1}{18 - 7} = \left(\frac{71 - 18}{t_m - 18} \right)^{1.3} = \frac{82 - 60}{t_2 - t_3}$$

Hence
$$\frac{19}{11} = \left(\frac{53}{t_m - 18}\right)^{1.3}$$

from which
$$t_m = 53°C$$

Also
$$\frac{19}{11} = \frac{22}{t_2 - t_3}$$

and
$$t_2 - t_3 = 12.7°C$$

therefore
$$t_2 = 53 + \frac{12.7}{2} = 59.4°C$$

$$t_3 = 53 - \frac{12.7}{2} = 46.7°C$$

Similarly,

(b) Convectors, for $t_m = 71°C$

$$\frac{18 + 1}{18 - 7} = \left(\frac{71 - 18}{t_m - 18}\right)^{1.5} = \frac{82 - 60}{t_2 - t_3}$$

Hence
$$\frac{19}{11} = \left(\frac{53}{t_m - 18}\right)^{1.5}$$

from which
$$t_m = 55°C$$

Since
$$t_2 - t_3 = 12.7°C$$

$$t_2 = 55 + \frac{12.7}{2} = 61.4°C$$

$$t_3 = 55 - \frac{12.7}{2} = 48.7°C$$

Comparison:

Temperature, °C

Inside	Outside	Radiators		Convectors	
		Flow	Return	Flow	Return
18	−1	82	60	82	60
18	7	59.4	46.7	61.4	48.7

This shows that if both radiators and convectors are used they should be supplied and controlled by separate water circuits.

Example 2.11. Calculate (a) the length of electric floor heating cable that will be required to heat the single-storey building specified below to the conditions stated and (b) the approximate cable spacing.

Data:

>   Wall area, 90 m^2; thermal transmittance = 1.7 W/m^2°C
>   Roof area, 30 m^2; thermal transmittance = 2.3 W/m^2°C
>   Roof glass area, 30 m^2; thermal transmittance = 6.8 W/m^2°C
>   Floor area, 50 m^2; surface temperature = 24°C
>   Internal surface resistance of wall and glass = 0.12 m^2°C/W
>   Internal surface resistance of roof = 0.10 m^2°C/W
>   Room air temperature: 16°C
>   External air temperature: -1°C
>   Electricity supply: 240 volt, unrestricted supply
>   Resistance of heating cable: 0.032 ohm/m
>   Downward and edge loss: 40 W/m^2

(*a*) Determine first the surface temperature of the unheated surfaces.

In general, from Eq (1.12) and Eq (1.13) we have:

$$t_{si} = t_i - R_{si} . U(t_i - t_o)$$

i.e.
$$t_{si} = 16 - R_{si} . U(16 + 1)$$
$$= 16 - 17 . R_{si} . U$$

then for Wall: $t_{si} = 16 - 17 \times 0.12 \times 1.7 = 12.5$°C

Glass: $t_{si} = 16 - 17 \times 0.12 \times 6.8 = 2.2$°C

Roof: $t_{si} = 16 - 17 \times 0.10 \times 2.3 = 12.1$°C

The area weighted mean temperature of the unheated surfaces (t_{mus}) will be:

$$t_{mus} = \frac{(90 \times 12.5) + (30 \times 2.2) + (30 \times 12.1)}{90 + 30 + 30} = 10\text{°C approx.}$$

Heat emission from the floor by thermal radiation will be from Eq (Int. 35):

$$\Phi_r/A = 4.93\left[\left(\frac{24 + 273}{100}\right)^4 - \left(\frac{10 + 273}{100}\right)^4\right] = 67.5 \text{ W/m}^2$$

and by convection from Eq (Int. 23), using C = 2.5 for heat flow upwards from the heated floor.

$$\Phi_c/A = 2.5(24 - 16)^{1.25} = 33.6 \text{ W/m}^2$$

The total upward emission from the floor is therefore 67.5 + 33.6 = 101 W/m^2 and the total installed load, including the downward and edge loss will be

$$101 + 40 = 141 \text{ W/m}^2$$

i.e.

$$141 \times 50 = 7\,050 \text{ W}$$

The total resistance of a single circuit of cable will be from Ohms $= \dfrac{\text{Volts}^2}{\text{Watts}}$:

$$\frac{240^2}{7\,050} = 8.17 \text{ ohms}$$

and the length of a single circuit of the given cable will be:

$$\frac{8.17}{0.032} = 255 \text{ m}$$

(b) Assuming that the entire floor area is heated then the cable spacing will be:

$$\frac{50}{255} = 20 \text{ cm approx.}$$

Example 3.12. A 2 m × 1 m sheet steel industrial type heating panel is fixed horizontally at high level and supplied with steam at 150°C.

(a) What thickness of insulation having a thermal conductivity of 0.043 W/m°C should be applied to the top side of the panel if the upper surface of the insulation is not to exceed 30°C.

(b) Calculate the total heat emission from the panel when: (i) insulated; (ii) uninsulated.

Use the following data:

    Ambient air temperature   18°C
    Mean temperature of surrounding surfaces   16°C

(a) Assuming steady heat flow, the rate of heat flow through the insulation must equal the total heat emission by radiation and convection from the upper surface of the insulation.

From Eq (Int. 4) the heat flow through the insulation is:

$$\Phi/A = \frac{0.043}{l}(150 - 30) = \frac{5.16}{l} \text{W/m}^2$$

The heat emission by radiation will be from Eq (Int. 35):

$$\Phi_r/A = 4.93\left[\left(\frac{30 + 273}{100}\right)^4 - \left(\frac{16 + 273}{100}\right)^4\right]$$
$$= 71.5 \text{ W/m}^2$$

and by convection from Eq (Int. 23); taking C = 2.5 for heat flow upwards:

$$\Phi_c/A = 2.5(30 - 18)^{1.25}$$
$$= 55.8 \text{ W/m}^2$$

Thus,

$$\frac{5.16}{l} = 71.5 + 55.8 = 127.3 \text{ W/m}^2$$

from which $l = \dfrac{5.16}{127.3} = 4{,}05$ cm.

(b) (i) Total emission from insulated panel

$$\text{Heat flow upwards} = 2 \times 127.3$$
$$= 254.6 \text{ W}$$

Heat flow downwards:

$$\text{by radiation} = 4.93 \left[\left(\frac{150 + 273}{100}\right)^4 - \left(\frac{16 + 273}{100}\right)^4 \right]$$
$$= 1\ 233 \text{ W/m}^2$$

and by convection (heat flow downwards):

$$1.3(150 - 18)^{1.25} = 581.6 \text{ W/m}^2$$

Total heat flow downwards:

$$2(1\ 233 + 581.6) = 3\ 629.2 \text{ W}$$

Total emission from insulated panel:

$$254.6 + 3\ 629.2 = 3\ 884 \text{ W}$$

(ii) Uninsulated panel

Heat flow upwards:

$$1\ 233 + 2.5(150 - 18)^{1.25} = 2\ 352 \text{ W/m}^2$$

or

$$2 \times 2\ 352 = 4\ 704 \text{ W}$$

Heat flow downwards, as in part (i) = 3 629 W. Total emission from uninsulated panel:

$$4\ 704 + 3\ 629 = 8\ 333 \text{ W}$$

Example 2.13. A room 6 m × 3 m × 3 m high is heated by a ceiling panel that covers the entire ceiling area. Consider an occupant seated at the centre of the room and determine the maximum temperature at which the panel may be operated for comfort. Assume that the mean radiant temperature at the head of the occupant should not exceed 20°C and that the mean temperature of the unheated surfaces in the room is 18°C.

Details of the panel are given in Fig. 2.7, which assumes that the head of the seated occupant may be represented by a small sphere with its centre

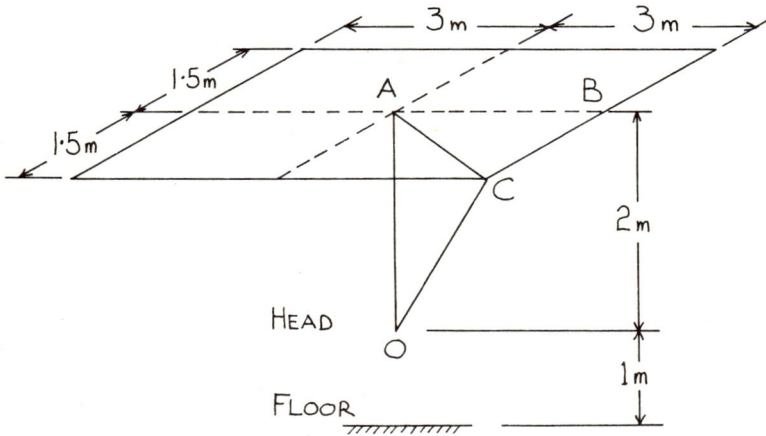

Fig. 2.7. Ceiling panel

1 m above the floor. The intensity of radiation from the panel is then a function of the solid angle subtended at 0. (A solid angle is by definition the ratio of the area of the surface of the portion of a sphere enclosed by the conical surface forming the angle, to the square of the radius of the sphere. The surface area of a sphere is $4\pi r^2$, and hence the solid angle subtended by the surface of the sphere at its centre is $\dfrac{4\pi r^2}{r^2} = 4\pi$. The unit of solid angle is the steradian which is that solid angle which encloses a surface on the sphere equal to the square of the radius. There are therefore 4π steradians in a sphere.) It may be shown that the solid angle subtended at 0 by the panel is:

$$\omega = 4\left(\tan^{-1}\frac{ab}{cd}\right) \text{ steradians}$$

where $a = AB$, $b = BC$, $c = AO$ and $d = CO$

$$AC = \frac{\sqrt{6^2 + 3^2}}{2} = 3.85 \text{ m}$$

$$CO = \sqrt{3.85^2 + 2^2} = 4.34 \text{ m}$$

Then
$$\omega = 4\left(\tan^{-1}\frac{3 \times 1.5}{2 \times 4.34}\right)$$

$$= 1.91 \text{ steradians}$$

The total solid angle at $O = 4\pi$ steradians, of which $\dfrac{1.91}{4\pi} = 0.152$ represents the panel and $1 - 0.152 = 0.848$ represents the remaining unheated surfaces of the room. Then, since the total energy emitted in the form of heat radiation from a surface is proportional to the fourth power of its absolute temperature, the panel surface temperature will be given by

$$T_p{}^4 = (20 + 273)^4 - 0.848(18 + 273)^4$$

From which

$$T_p = 307 \text{ K}$$

and

$$t_p = 307 - 273 = 34°\text{C}$$

Example 2.14. A room which has a solid ground floor, measuring 15 m × 8 m and having four exposed edges, is to be heated by 20 mm O.D. hot-water pipes embedded in the floor at 230 mm centres and 75 mm to their centre below the floor surface. Using the data listed below, calculate:

(*a*) the upward heat emission from the surface of the floor to the room;
(*b*) the approximate downward heat loss from the heating pipes;
(*c*) the total heat emission from the pipes;
(*d*) the water flow rate and flow and return water temperatures.

Data:

Floor surface temperature 24°C
Room air temperature 16°C
Outside air temperature −1°C
Coefficient of heat transfer at floor surface = 9.65 W/m²°C
Thermal conductivity of floor material = 1.44 W/m°C
Temperature drop in water circuit = 10°C
Specific heat capacity of water = 4.2 kJ/kg°C

(*a*) Total upward heat emission

$$15 \times 8 \times 9.65(24 - 16) = 9\ 264 \text{ W}$$

(*b*) The conventional U value for a solid floor 15 m × 8 m with four exposed edges is found from the IHVE Guide to be 0.625 W/m²°C, which includes the effect of the normal surface resistance to heat flow downwards, and which would normally be used in conjunction with the design inside and outside air temperature. In this example we are concerned with the approximate heat flow downwards from the level of the heating pipes. It will be necessary therefore to correct the conventional U value so that it may be used in conjunction with the mean temperature at the level of the pipes and the outside air temperature. This may be done by deducting from the total resistance equivalent to the conventional U value, the resistance at the surface for heat flow down and the resistance of the floor slab above

the pipes. Since the surface resistance is 0.15 m²°C/W for heat flow downwards the corrected U value will be:

$$(U)_{corrected} = \left(\frac{1}{0.625} - 0.15 - \frac{0.075}{1.44}\right)^{-1}$$
$$= 0.715 \text{ W/m}^2\text{°C}$$

Since the rate of heat flow from the surface is equal to the rate of heat
flow from the pipes through the floor slab to the floor surface, we have:

$$\frac{t_p - 24}{\dfrac{0.075}{1.44}} = 9.65(24 - 16)$$

where t_p = mean temperature at pipe level,

i.e.
$$t_p = 24 + \frac{0.075}{1.44} \times 9.65(24 - 16)$$
$$= 28\text{°C}$$

The approximate downward heat loss is therefore:

$$15 \times 8 \times 0.715(28 + 1) = 2\,488 \text{ W}$$

(c) The total emission from the pipe is:

$$9\,264 + 2\,488 = 11\,752 \text{ W say } 11.75 \text{ kW}$$

The downward heat loss in this example is

$$\frac{2\,488}{11\,752} \times 100 = 21 \text{ per cent approx. of the total emission}$$

or $\dfrac{2\,488}{9\,264} \times 100 = 27$ per cent approx. of the upward emission from the
floor surface

(d) The total water flow rate for the installation will be:

$$\frac{11.75}{10 \times 4.2} = 0.28 \text{ kg/s}$$

The mean temperature of the water in the pipes may be determined only
approximately by using the mean resistance of the floor above the pipes
and the rate of heat flow upwards.

The mean resistance $= \dfrac{a + b}{2\lambda}$

where

a = depth of cover directly above the pipe surface to the floor surface,

i.e. $75 - \dfrac{20}{2} = 65 \, mm$

b = distance between the pipe surface and the floor surface half-way

between two pipes, i.e. $\sqrt{75^2 + 115^2} - \dfrac{20}{2} = 127 \, mm$

Then, the mean water temperature (t_w) will be:

$$t_w = 24 + \left(\dfrac{0.065 + 0.127}{2 \times 1.44}\right) \times 9.65(24 - 16)$$

$$= 29°C \text{ approx.}$$

Since the temperature drop in the water circuit is to be 10°C, the flow and return water temperatures will therefore be 34°C and 24°C respectively.

Example 2.15. A unit heater has a heat emission of 15 kW with air entering at 24°C and leaving at 46°C when supplied with steam at 150°C. What will be the heat emission if the steam temperature is reduced to 110°C and the inlet air is 16°C?

It was shown in the section on unit heaters that:

$$\Phi = UA \, \dfrac{t_2 - t_1}{\log_e \dfrac{t_s - t_1}{t_s - t_2}}$$

Then

$$\dfrac{\Phi_{I}}{\Phi_{II}} = \dfrac{(t_2 - t_1)_{I}}{(t_2 - t_1)_{II}} \, \dfrac{\log_e \left[\dfrac{t_s - t_1}{t_s - t_2}\right]_{II}}{\log_e \left[\dfrac{t_s - t_1}{t_s - t_2}\right]_{I}}$$

Where subscripts I and II refer to the initial and final conditions. It was also shown earlier that $\Phi \propto (t_2 - t_1)$ therefore:

$$\Phi_{I}/\Phi_{II} = (t_2 - t_1)_{I}/(t_2 - t_1)_{II}$$

thus

$$[(t_s - t_1)/(t_s - t_2)]_{II} = [(t_s - t_1)/(t_s - t_2)]_{I}$$

i.e.

$$\dfrac{110 - 16}{110 - t_2} = \dfrac{150 - 24}{150 - 46}$$

from which

$$t_2 = 37.5°C$$

The new heat emission will be; since $\Phi \propto (t_2 - t_1)$

$$15 \times \frac{37.5 - 16}{46 - 24} = 14.65 \text{ kW}$$

Example 2.16. If in the above example the fan speed was 16 rev/s what will be the heat emission with air entering at 18°C, steam at 120°C and a fan speed of 12 rev/s?

It was shown earlier that $\Phi = UA\Delta t_m = \rho u_1 c(t_2 - t_1)$ and that t $U \propto u_1{}^{0.6}$ and also that $u_1 \propto N$. Then by appropriate ratios and substitutions it may be shown that:

$$\left(\frac{N_{II}}{N_I}\right)^{0.4} = \log_e\left[\frac{t_s - t_1}{t_s - t_2}\right]_I \div \log_e\left[\frac{t_s - t_1}{t_s - t_2}\right]_{II}$$

i.e.

$$\left(\frac{12}{16}\right)^{0.4} = \log_e\left[\frac{150 - 24}{150 - 46}\right] \div \log_e\left[\frac{120 - 18}{120 - t_2}\right]$$

from which

$$t_2 = 38°C$$

The new heat emission will be

$$15 \times \frac{12}{16} \times \frac{38 - 18}{46 - 24} = 10.2 \text{ kW.}$$

Problems

1. A pipe of 76 mm outside diameter conveys steam at 204°C and is insulated with a layer of magnesia 51 mm thick for which the thermal conductivity is 0.07 W/m°C. The room temperature is 15.6°C and the coefficients of heat transfer at the inside of the pipe and the outside of the insulation are 852 and 11.4 W/m²°C respectively. Neglect thickness and resistance of pipe wall and calculate:

(*a*) the rate of heat loss from the insulated pipe;
(*b*) the outside surface temperature of the insulation;
(*c*) the efficiency of the insulation if the emission of the uninsulated pipe is 865 W/m   (NC)

Ans.: (*a*) 90 W/m approx.; (*b*) 29.4°C; (*c*) 90 per cent. (approx.)

2. A two-pipe forced-circulation low-pressure hot-water heating system is designed to maintain the air temperature in a building at 18.3°C when the outside air temperature is −1.1°C. If the system design flow and return water temperatures are 76.7°C and 65.6°C respectively, determine the system flow and return water temperatures required to maintain the inside air temperature at 18.3°C when the outside air temperature is 10°C for:

(*a*) radiator heating appliances;
(*b*) natural convector heating appliances.

Neglect heat losses from the pipe work and assume constant thermal transmittance and rate of air change.   (NC)

Ans.: (*a*) 48.3°C and 43.3°C; (*b*) 50.6°C and 46.7°C. (approx.)

3. A room is maintained at 18.3°C when −1.1°C outside by means of hot-water radiators having a mean surface temperature of 71.1°C. During a test the outside temperature remained constant at 5.6°C and the mean surface temperature of the radiators was constant at 71.1°C. What temperature would you expect in the room?   (NC)

Ans.: 22.8°C.

4. A 114 mm O.D. horizontal pipe is insulated with 64 mm thickness of 85 per cent magnesia having a thermal conductivity of 0.065 W/m°C. The temperature of the air and surroundings are uniformly at 18.3°C. Calculate the heat emission per square metre of external surface if the pipe carries steam at a temperature of 121°C. Assume the surface of the insulation has an emissivity of 0.9. Neglect thickness and resistances of the pipe wall and the resistance at the inside surface.

Ans.: 68 W/m^2; t_{So} = 27°C. approx.

5. A boiler plant operates in such a manner as to provide a constant-flow temperature of 82.2°C throughout the heating season. A circulation is taken from the boiler headers through a three-way mixing valve to feed cast-iron radiators proportioned to provide for a building internal temperature of 15.6°C and −1.1°C outside. The mixing valve is controlled by an external pilot to provide a circuit flow of 71.1°C under design conditions and to vary that temperature with rise in external temperature. The design temperature drop across the circuit is 11°C.

If the building heat requirement varies directly with changes in outside temperature, calculate the proportions of boiler and by-pass water which flow through the mixing valve for outside temperatures of −1.1°C, 4.4°C and 10°C. The effect of mains emission may be ignored.   (IHVE)

Ans.: −1.1°C : 0.5 and 0.5. 4.4°C : 0.2 and 0.8. 10°C : 0.07 and 0.93.

6. A heating system in a warehouse consists of exposed 80 mm-bore water piping which requires a mean water temperature of 71.1°C to maintain 15.6°C in the space when −1.1°C outside.

Due to structural alterations, the building heat loss per degree difference is subsequently reduced to 90 per cent of its former value. What percentage of the piping must be insulated (to an efficiency of 76 per cent) so as to maintain a temperature of 12.8°C in the space when −1.1°C outside, the mean water temperature remaining the same?

Heat loss from piping to be taken as proportional to the temperature difference to the power of 1.3. Loss from 80 mm pipe at 66°C difference is 244 W/m.   (IHVE)

Ans.: 39 per cent. approx.

7. The total heat emission from the radiators in a room maintained at 18.3°C is 29.3 kW when the mean radiator surface temperature is 71.1°C. What will be the radiators emission if the mean surface temperature drops to 65.6°C. If convectors are used instead of radiators, what will be the variation in output?   (NC)

Ans.: New emission; radiators 25.4 kW, convectors 24.8 kW.

8. A building is maintained at 15.6°C when −1.1°C outside by a hot-water radiator heating system operating with flow and return water temperatures of 71.1°C and 60°C respectively. The boiler also supplies an indirect hot-water supply system and is controlled to operate at a constant flow-water temperature of 82.2°C. Draw a diagram of the arrangement and calculate the proportions of boiler water and return water that must pass through the mixing valve for outside temperatures of −1.1°C and 4.4°C.   (NC)

Ans.: −1.1°C, 1 : 1; 4.4°C, 1 : 4.

9. A building, heated by means of a two-pipe forced-circulation low-pressure hot-water radiator heating system, contains one room (A) which measures 12.2 m by 9.1 m by 4.2 m high. Room (A) has two external walls facing N and E constructed of 152 mm concrete, an air gap and 19 mm plasterboard, 45 per cent of the wall area being glass.

The heating system is designed for flow and return water temperatures of 76.7°C and 65.6°C respectively and an outside air temperature of −1.1°C. A mixing valve is used for control of the heat output.

If the design inside air temperature of room (A) is 18.3°C, neglecting pipe and partition heat losses, determine from the data given:

(a) The steady state heat requirement of room (A).

(b) The area of radiator heating surface required for room (A).

(c) The radiator flow and return water temperatures required to maintain room (A) at 7.2°C air temperature when −1.1°C outside air temperature and steady state conditions exist. Assume constant thermal transmittance and rate of air change.

Data:

Thermal transmittance of glass	= 5.678 W/m²°C
Thermal transmittance of a 6-in concrete building wall	= 3.407 " " "
Thermal transmittance of roof of room (A)	= 1.42 " " "
Thermal transmittance of floor of room (A)	= 0.568 " " "
Thermal resistivity of plasterboard	= 17.3 m°C/W
Thermal resistance of air space	= 0.176 m²°C/W
Air change for room (A)	= 1.5/h
Radiator emission	= 600 W/m² for 60°C temperature difference (NC)

Ans.: (*a*) 13.7 kW; (*b*) 90 m; (*c*) 37°C and 32°C. approx.

10. A factory building having a heat loss of 220 kW is maintained at 15.6°C when −1.1°C outside by hot-water radiators having a mean water temperature of 71.1°C and an emission of 600 W/m² 60°C. Subsequently, it is desired to maintain the building at 23.9°C when −1.1°C outside, and to meet this requirement it is proposed to increase the mean water temperature of the radiator to 76.7°C. How much additional radiator heating surface should be installed?   (NC)

Ans.: 245 m². approx.

11. A 2.44 m × 1.22 m sheet steel industrial type heating panel is fixed horizontally at high level and supplied with hot water having a mean temperature of 149°C.

(*a*) What thickness of insulation having a thermal conductivity of 0.043 W/m°C should be applied to the top side of the panel if the upper surface of the insulation is not to exceed 37.8°C.

(*b*) Calculate the total heat emission from the panel when: (i) insulated, and (ii) uninsulated.

Use the following data:

Ambient air temperature	15.6°C
Mean temperature of surrounding surfaces	12.8°C (NC)

Ans.: (*a*) 18 mm; (*b*) (i) 6.5 kW; (ii) 13.1 kW.

3: Hot-Water Heating Systems

General Considerations

Circulating Pressure

The circulating pressure or motive force in a natural circulation hot-water heating system is porportional to the difference in temperature, and hence the difference in density, of the ascending and descending vertical columns of water and their height. Consider the simple circuit shown in Fig. 3.1 in which:

t_1 = mean temperature of the ascending hot flow water
t_2 = mean temperature of the descending cooler return water
ρ_1 = density of water at temperature t_1
ρ_2 = density of water at temperature t_2

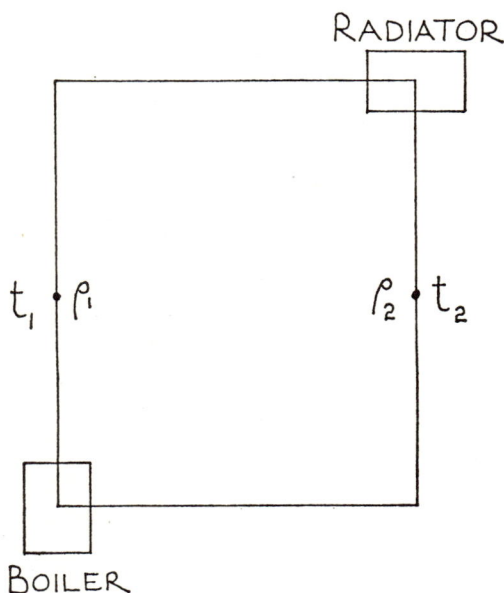

Fig. 3.1. Circulating pressure, basic arrangement.

Since $t_1 > t_2, \rho_2 > \rho_1$ and the pressure difference (Δp) across the circuit will be:

$$\Delta p = hg(\rho_2 - \rho_1) \tag{3.1}$$

where h = circuit height,
and g = standard gravity.

TABLE 3.1

t_1	*Temperature difference $t_1 - t_2$ and circulating pressure Δp, N/m^2 per metre of circuit height h. (approx.)*					
	5	10	15	20	25	30
70	20	50	–	–	–	–
75	30	50	80	–	–	–
80	30	60	80	110	–	–
85	30	60	90	110	140	–
90	40	70	100	130	150	180

The amount of natural circulating pressure is very small, see Table 3.1, and leads to comparatively large diameters. While, for this reason, modern systems are designed for forced circulation, it may be necessary to consider the natural circulating effect when sizing connections to units served from one-pipe systems and in general the amount of pressure available will be:

$$\Delta p = \Delta p_A + \Delta p_B \tag{3.2}$$

where

Δp_A = pressure drop along main between connections to the unit,

Δp_B = natural circulating pressure in the sub-circuit, being positive when above main and negative when below.

In general, the circuit height (h) is considered to be positive when measured above the thermal centre of the heat source and negative when measured below.

For natural-circulation systems with high level or dipped returns the circulating pressure is therefore found from the algebraic sum of the positive and negative circulating effects.

With a one-pipe drop system as shown in Fig. 3.3 the temperature distribution, and hence the water density, changes throughout the system are not normally known at the outset of a design. It is necessary therefore to use a mean effective height for the dropper when making a preliminary estimate of the circulating pressure. For the dropper shown in Fig. 3.3, and neglecting the effect of the cooling in the horizontal mains, the mean effective height (h_m) will be:

$$h_m = \frac{h_1 \cdot \Phi_1 + h_2 \cdot \Phi_2 + h_3 \cdot \Phi_3}{\Phi_1 + \Phi_2 + \Phi_3} \tag{3.3}$$

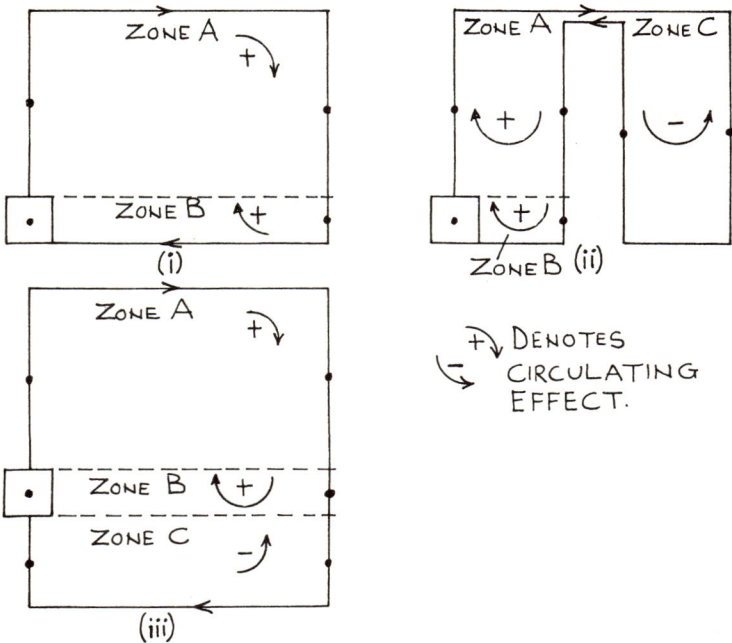

Fig. 3.2.

where h_m = mean effective height of dropper
Φ_{1-3} = heat loads
h_{1-3} = height.

Temperature Drop across Circuit

The difference between the flow and return water temperature should be fixed with respect to the required mean temperature of the heating unit, the mode of water circulation and the type of heat source. Typical temperature drops are given in Table 3.1.

TABLE 3.1

System	Design flow water temp., $^{\circ}C$	Max. temp. drop across circuit, $^{\circ}C$
Embedded ceiling panel coils	55	10
Embedded floor panel coils	45	10
Embedded sleeved panel coils	70	10
Radiators, convectors, unit heaters, low-temp. metal plate panels and skirting heaters (forced circulation)	70 to 90	10 to 20
Radiators (natural circulation)	80	20

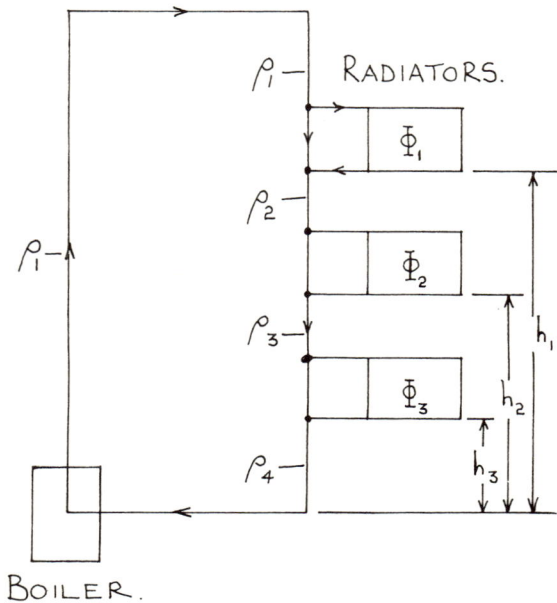

Fig. 3.3.

In the case where several heat loads are served by the same dropper of a one-pipe system, as in Fig. 3.3, there exists a definite relationship between the temperature drop across the radiator connections and the temperature drop across the dropper. The design temperature drop across any heat load served by the dropper should be at least $2°C$ above the minimum value, i.e.

$$\Delta t = (\Phi/\dot{m} . c) + 2 \qquad (3.4)$$

where

Δt = minimum temperature drop across heat load
Φ = heat load
$\dot{m}$ = initial flow rate in dropper.
c = specific heat capacity.

Water Flow Rate

The temperature drop from Table 3.1 or Eq (3.4) is used to obtain the required water flow rate:

$$\dot{m} = \frac{\Sigma\Phi}{\Delta t . c} \qquad (3.5)$$

where

   $\dot{m}$ = water flow rate

   $\Sigma\Phi$ = total heat load on system from all heating units and attendant pipework

   Δt = difference in temperature between flow and return water at the boiler

   c = specific heat capacity.

It should be noted that the total heat load ($\Sigma\Phi$) for any hot-water system is not known until the pipe sizes have been fixed. For this reason it is not possible to obtain an accurate estimate of the final water flow rates, and a trial-and-error method of solution cannot be avoided. An arbitrary 30 per cent is generally added to the net heat loads when making preliminary estimates of the flow rate. A subsequent critical analysis of the preliminary design calculations will give the true relationship between pipe and unit emissions.

Index Circuit

Consider the simple two-circuit arrangement shown in Fig. 3.4 and assume that the temperature drop across each circuit is the same. Then, since the height of A is twice the height of B, the total circulating pressure for the right-hand circuit will be twice that for the left-hand circuit. Pipes *a* and *b* are common to both circuits and must be capable of passing the combined flow rate of A and B under the least-favourable pressure conditions. The pressure available per unit length will be least for the circuit to B, which has the greatest length and the smallest total circulating pressure. Unit B is referred to as the index unit, and its circuit is called the index circuit. For natural circulation systems the index circuit is that circuit for which the ratio of circuit height to circuit length is a minimum. With forced-circulation systems the natural circulating effect is generally negligible and the index circuit for a fixed pump head is always taken to be the longest circuit.

Methods of Proportioning Mains Emission

The additional amount of water required to be circulated through a system due to the heat emission from the pipes should be carried in some proportion by the various branches. This may be carried out in several ways depending on the data that are required in the answer. The methods commonly used are given below and illustrated in Example (3.4) that follows:

(*a*) By calculation of the temperature drop along each section of the pipework, and hence the temperature drop across each branch. This method yields full information on flow rate, temperature distribution and mean temperature of each unit. The time involved in this procedure is greater

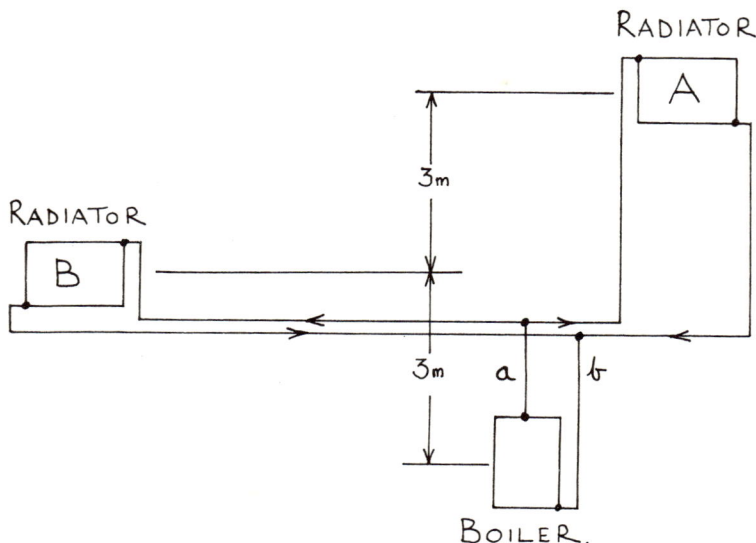

Fig. 3.4

than the time taken to solve the same problem by the alternative methods that follow. The process may be speeded up by tabulating the calculations.

(*b*) By direct ratio methods carried out either on the plans or by tabulated calculations on separate sheets. Some prefer to work from the heat source forward to the index unit, while others work back from the index unit to the heat source. The tabulation method generally takes longer than working directly on the plans. This method gives the flow rates only. If a temperature distribution is required further calculations have to be carried out.

(*c*) By graphical representation.* This method is suitable for illustrating the problem of mains proportioning. It is unlikely that it would be used in preference to either (*a*) or (*b*). With this method the greatest difficulty is in selecting a suitable scale to show clearly both large and small heat loads. The flow rate only results from this method.

The basic construction for this method is shown in Fig. 3.5, in which, drawn to a suitable scale:

$$AO = BO = \Sigma\Phi$$

$\Sigma\Phi$ = total heat load. Units plus pipe emission, W

---

* See also FABER, O., and KELL, J. R., *Heating and Air Conditioning of Buildings*, p. 208, 3rd Edition, Architectural Press.

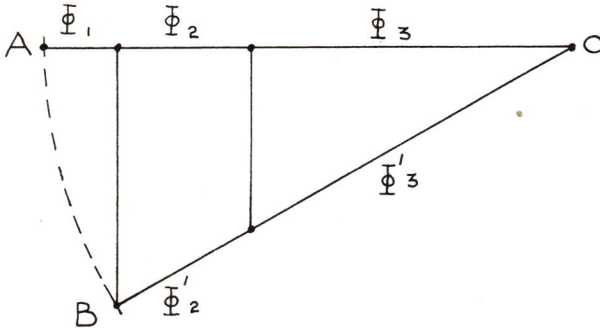

Fig. 3.5. Graphical representation of pipe emission.

Φ_1 = heat emission from mains serving branch loads Φ_2 and Φ_2, W

Φ_2 and Φ_3 = branch loads before proportioning, W

Φ_2' and Φ_3' = branch loads after proportioning, W

$\Phi_2' + \Phi_3' = \Sigma\Phi$.

Then since

$$\frac{\Phi_2}{\Phi_2 + \Phi_3} = \frac{\Phi_2'}{\Phi_2' + \Phi_3'}$$

$$\Phi_2' = \frac{\Phi_2}{\Phi_2 + \Phi_3} \cdot \Sigma\Phi$$

Example 3.1. Determine the preliminary circulating pressure and pressure drop for each branch of the two-pipe up-feed system shown in Fig. 3.6. Take the mean flow and return water densities at 972 kg/m³ and 983 kg/m³ respectively and assume that the lengths given include an allowance for local resistances.

For A and C, which have a circuit height of 2 m, the circulating pressure will be, from Eq (3.1):

$$\Delta p = 2 \times 9.81(983 - 972) = 216 \text{ N/m}^2$$

and for B and D, which have a circuit height of 5 m:

$$\Delta p = 5 \times 9.81(983 - 972) = 540 \text{ N/m}^2$$

For the purpose of finding the pressure drop for each branch consider the arrangement shown in Fig. 3.7, which is essentially a parallel representation of the four circuits of Fig. 3.6.

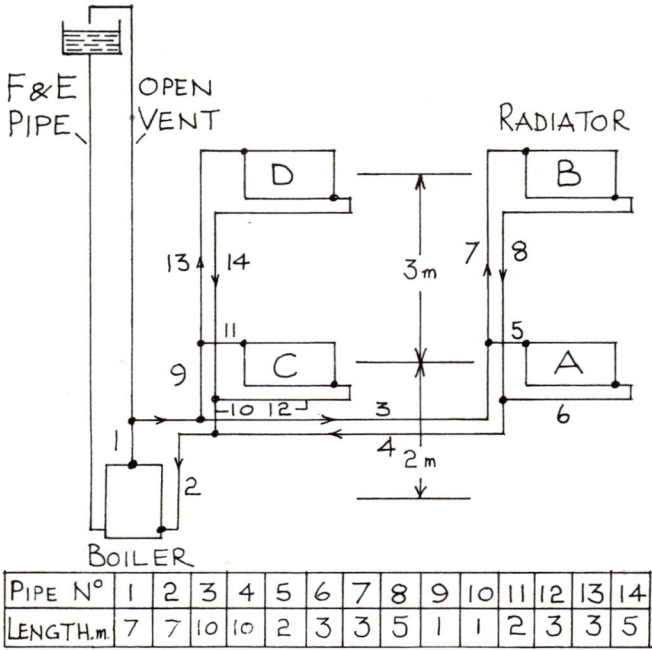

PIPE Nº	1	2	3	4	5	6	7	8	9	10	11	12	13	14
LENGTH.m.	7	7	10	10	2	3	3	5	1	1	2	3	3	5

Fig. 3.6. Two-pipe up-feed system.

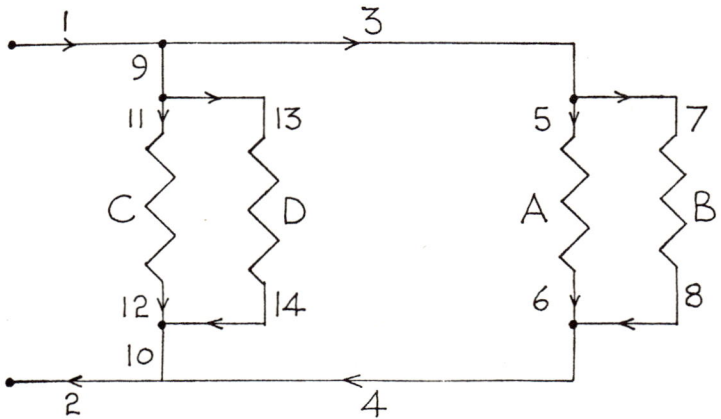

Fig. 3.7. Parallel circuits.

By inspection, Fig. 3.6, A is the index radiator and its circuit, pipes 1—6, is the index circuit. The preliminary pressure drop for this circuit will be:

$$\frac{\Delta p}{l} = \frac{216}{39} = 5.54 \; \frac{N}{m^2} \Big/ m$$

The circuit to B consists of pipes 1, 2, 3, 4, 7 and 8, of which pipes 1—4 are common to the index circuit and will be sized to absorb pressure at the rate of $5.54 \frac{N}{m^2} \Big/ m$.

The pressure drop of pipes 1—4 will be

$$5.54 \times 34 = 188 \; \frac{N}{m^2}$$

Therefore of the total circulating pressure of 540 N/m² available for the circuit to B:

$$540 - 188 = 352 \; N/m^2$$

will be available for sizing pipes 7 and 8. The pressure drop for these pipes will then be:

$$\frac{352}{8} = 44 \; \frac{N}{m^2} \Big/ m$$

The circuit to C consists of pipes 1, 2, 9, 10, 11 and 12, of which 1 and 2 are common to the index circuit and have a pressure drop of:

$$5.54 \times 14 = 78 \; \frac{N}{m^2}$$

Therefore of the total of $216 \frac{N}{m^2}$ available for C:

$$216 - 78 = 138 \; \frac{N}{m^2}$$

will be available for sizing pipes 9—12. The pressure drop for these pipes will be:

$$\frac{138}{7} = 19.7 \; \frac{N}{m^2} \Big/ m$$

The circuit to D contains pipes 1 and 2, which are common with the index circuit, and pipes 9 and 10, which are common with the circuit to C. The pressure drop for 13 and 14 will then be:

$$540 - 78 - 19.7 \times 2 = 423 \; \frac{N}{m^2}$$

i.e.

$$\frac{423}{8} = 52.9 \; \frac{N}{m^2} \Big/ m$$

Summary:

A preliminary pipe sizing may now be carried out using the following pressure drops:

Pipe No.	Preliminary pressure drop, $\frac{N}{m^2}\Big/m$
1 to 6	5.54
7 and 8	44
9 to 12	19.7
13 and 14	52.9

Example 3.2. Determine the flow rate (kg/s) for each section of the hot-water piping layout shown in Fig. 3.8 if the specific heat capacity of the water is 4.2 kJ/kg°C.

Alternative solutions are given, based on the methods of proportioning pipe emission outlined earlier.

Method (*a*). By calculation of temperature drop.

In general: $\qquad\qquad \Phi = \dot{m}.c.\Delta t$

and $\qquad\qquad\qquad \Delta t = \Phi/\dot{m}.c$

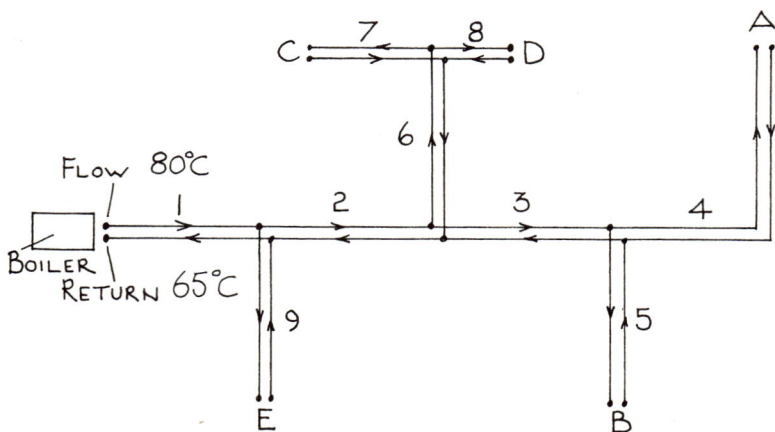

UNIT	A	B	C	D	E				
HEAT EMISSION kW	20	15	10	5	15				
PIPE SECTION (F+R)	1	2	3	4	5	6	7	8	9
HEAT EMISSION kW	4	1	1	5	2	2	2	1	2

Fig. 3.8. Two-pipe system.

where Φ = heat flow
$\dot{m}$ = mass flow
c = specific heat capacity
Δt = temperature difference.

Let subscripts 1, 2, ... 9 refer to the respective pipe sections.

Heat emission from all units = 65
Heat emission from all pipes = 20

Total heat load = 85 kW

$$\therefore \dot{m}_1 = \frac{85}{4.2(80-65)} = 1.35 \text{ kg/s}$$

and

$$\Delta t_1 = \frac{4}{1.35 \times 4.2} = 0.7°C$$

$\therefore$ temperature drop across branch to unit E and Section 2 will be $15 - 0.7 = 14.3°C$.
Total heat load on section 9 = 15 + 2 = 17 kW

$$\therefore \quad \dot{m}_9 = \frac{17}{14.3 \times 4.2} = 0.283 \text{ kg/s}$$

and, by difference

$$\dot{m}_2 = 1.35 - 0.283 = 1.067 \text{ kg/s}$$

$$\Delta t_2 = \frac{1}{1.067 \times 4.2} = 0.22°C$$

$\therefore$ temperature drop across branch to section 6 and section 3 will be $14.3 - 0.22 = 14.08°C$.
Total heat load on section 6 = 2 + 10 + 2 + 5 + 1 = 20 kW

$$\therefore \quad \dot{m}_6 = \frac{20}{14.08 \times 4.2} = 0.338 \text{ kg/s}$$

and

$$\dot{m}_3 = 1.067 - 0.338 = 0.729 \text{ kg/s}$$

$$\Delta t_3 = \frac{1}{0.729 \times 4.2} = 0.33°C$$

$\therefore$ temperature drop across branch to section 4 and section 5 will be $14.08 - 0.33 = 13.75°C$.

Total heat load on section 5 = 15 + 2 = 17 kW.

$$\therefore \quad \dot{m}_5 = \frac{17}{13.75 \times 4.2} = 0.294 \text{ kg/s}$$

and $$\dot{m}_4 = 0.729 - 0.294 = 0.435 \text{ kg/s}$$

$$\Delta t_6 = \frac{2}{0.338 \times 4.2} = 1.41°C$$

∴ temperature drop across branch to section 7 and 8 will be $14.08 - 1.41 = 12.67°C$.

Total heat load on section 7 = 10 + 2 = 12 kW

∴ $$\dot{m}_7 = \frac{12}{12.67 \times 4.2} = 0.226 \text{ kg/s}$$

and $$\dot{m}_8 = 0.338 - 0.226 = 0.112 \text{ kg/s}$$

Section No.	1	2	3	4	5	6	7	8	9
$\dot{m}$, kg/s ...	1.35	1.067	0.729	0.435	0.294	0.338	0.226	0.112	0.283

Method (*b*). By direct ratio.

(i) Working forward from the boiler:

$$\dot{m}_1 = 1.35 \text{ kg/s}$$

The heat emission from section 1, 4 kW, represents a flow rate of

$$\frac{4}{4.2(80 - 65)} = 0.063 \text{ kg/s}$$

this will be divided between section 2 and section 9 in direct ratio to the net loads carried by these branches as follows:

Section 2: Total heat load = 64 kW

Section 9: Total heat load = 17 kW

Proportion to section $2 = \dfrac{64 \times 0.063}{64 + 17} = 0.05 \text{ kg/s}$

∴ Proportion to section 9 = 0.063 − 0.05 = 0.013 kg/s

Net flow rate for section 9 and unit E $= \dfrac{17}{4.2 \times 15} = 0.27 \text{ kg/s}$

Proportion from section 1   = 0.013 kg/s

$$\dot{m}_9 = \overline{0.283} \text{ kg/s}$$

∴ $$\dot{m}_2 = 1.35 - 0.283 = 1.067 \text{ kg/s}$$

This checks with the flow rate for section 2 determined by method (*a*). Similar calculations may be carried out for the remaining sections.

(ii) By direct ratio on the diagram. The heat loads would be accumulated for each section as shown in Fig. 3.9. All calculations would then be carried out by slide-rule, but for completeness the calculations involved are shown below:

$$\dot{m}_1 = \frac{85}{4.2(80 - 65)} = 1.35 \text{ kg/s}$$

$$\dot{m}_2 = \frac{64}{81} \times 1.35 = 1.067 \text{ kg/s}$$

$$\dot{m}_9 = 1.35 - 1.067 = 0.283 \text{ kg/s}$$

$$\dot{m}_3 = \frac{43}{63} \times 1.067 = 0.729 \quad \text{''}$$

$$\dot{m}_6 = 1.067 - 0.729 = 0.338 \text{ ''}$$

$$\dot{m}_4 = \frac{25}{42} \times 0.729 = 0.435 \quad \text{''}$$

$$\dot{m}_5 = 0.729 - 0.435 = 0.294 \text{ ''}$$

$$\dot{m}_7 = \frac{12}{18} \times 0.338 = 0.226 \quad \text{''}$$

$$\dot{m}_8 = 0.338 - 0.226 = 0.112 \text{ ''}$$

Method (c). By graphical representation.

The general construction is shown in Fig. 3.10, which for clarity of presentation is not drawn to scale.

Line AE represents the total heat load   = 85 kW
Line AB represents mains loss, section 1   =   4 ''
Line BB′ represents load on section 9   = 17 ''
Line B′C represents mains loss, section 2 =   1 ''
Line CC′ represents load on section 6   = 20 ''
Line C′D represents mains loss, section 3 =   1 ''
Line DD′ represents load on section 5   = 17 ''
Line D′E represents load on section 4   = 25 ''
Line 2F represents mains loss, section 6   =   2 ''
Line FG represents load on section 7   = 12 ''
Line 1E represents the total heat load   = 85 ''

1E is drawn at any convenient angle to AE.
B′2 is drawn parallel to B1.
C′3 is drawn parallel to C2.
D′4 is drawn parallel to D3.
G6 is drawn parallel to F5.

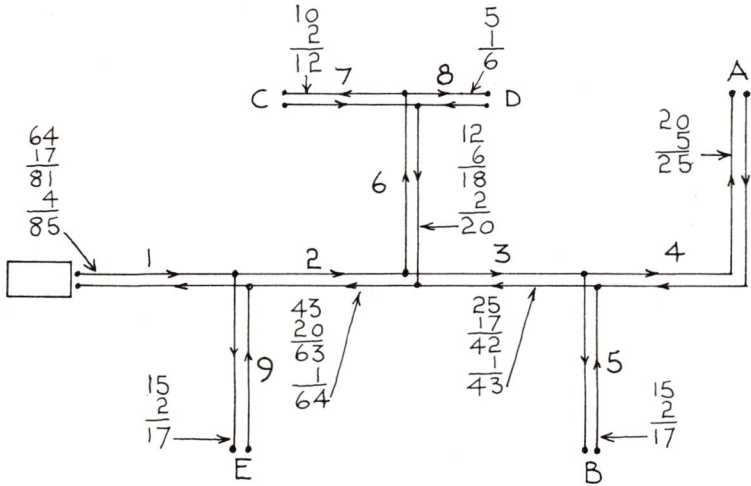

Fig. 3.9. Load proportioning.

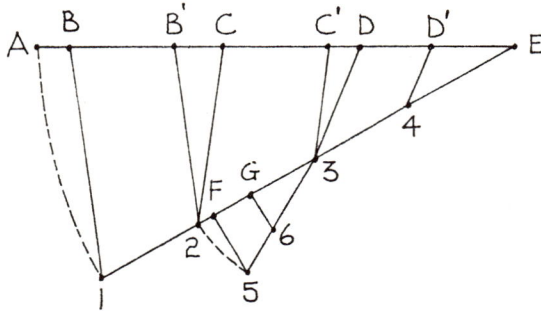

Fig. 3.10. Graphical solution of load proportioning.

When the diagram is drawn to scale the proportional heat load for each section may be scaled off as follows:

Line 1−2 = heat load on section 9
Line 2−3 = heat load on section 6
Line 3−4 = heat load on section 5
Line 4−5 = heat load on section 4
Line 5−6 = heat load on section 7
Line 6−3 = heat load on section 8

In each case these heat loads include the net heat load for the section plus a proportional share of the heat emission from the other sections. The flow rates are obtained by dividing the above heat loads by the temperature drop for the circuit and the specific heat capacity. Consider the load on branch 9 and refer to the earlier description of the graphical method.

$$BB' = 17 \text{ kW}$$
$$B'E = 85 - 17 - 4 = 64 \text{ kW}$$
$$BE = 64 + 17 = 81 \text{ kW}$$
$$1E = 85 \text{ kW}$$

Then
$$1-2 = \frac{17}{81} \times 85 = 17.85 \text{ kW}$$

and
$$\dot{m}_9 = \frac{17.85}{4.2(80 - 65)} = 0.283 \text{ kg/s}$$

The flow rates by all methods are thus:

Section No.	1	2	3	4	5	6	7	8	9
$\dot{m}$, kg/s	1.35	1.067	0.729	0.435	0.294	0.338	0.226	0.112	0.283

Example 3.3. Determine for the hot-water heating system shown in Fig. 3.11:

(*a*) the water flow rate (kg/s) through each section;
(*b*) the mean temperature of each radiator.

Assume that the specific heat capacity of the water is 4.2 kJ/kg°C.

Total heat emission from radiators = 19 kW
Total heat emission from pipes   =   3.6 ,,
                                  ——————
                                    22.6 ,,

$$\dot{m}_{1,2} = \frac{22.6}{4.2(85 - 60)} = 0.215 \text{ kg/s}$$

$$\Delta t_1 = \frac{0.3}{0.215 \times 4.2} = 0.33°C$$

Initial temperature of pipe 1 = 85°C
Final temperature of pipe 1 = 85 - 0.33 = 84.67°C

$$\Delta t_2 = \frac{0.2}{0.215 \times 4.2} = 0.22°C$$

Initial temperature of pipe 2 = 60°C
Final temperature of pipe 2 = 60 + 0.22 = 60.22°C

∴ temperature drop across branch 17 and 18 and branch 3 and 4 = 84.67 - 60.22 = 24.45°C.

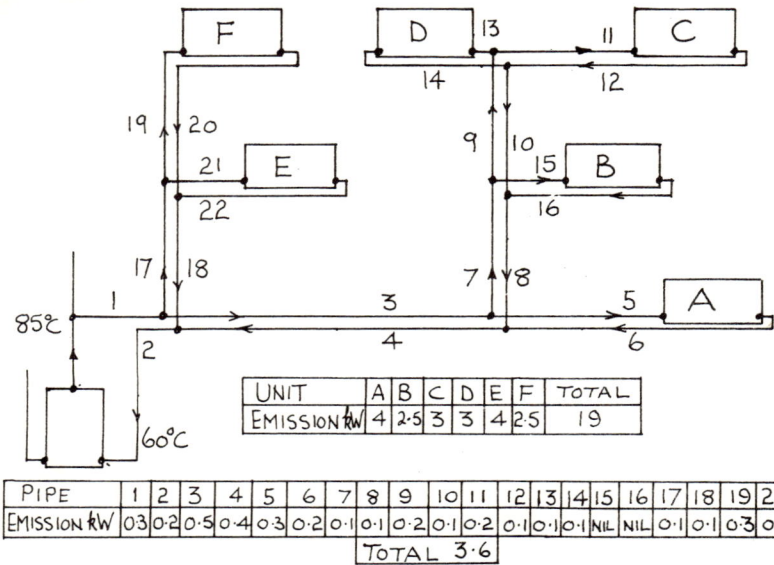

Fig. 3.11. Two-pipe system, mains emission.

UNIT	A	B	C	D	E	F	TOTAL
EMISSION kW	4	2.5	3	3	4	2.5	19

PIPE	1	2	3	4	5	6	7	8	9	10	11	12	13	14	15	16	17	18	19	20	21	22
EMISSION kW	0.3	0.2	0.5	0.4	0.3	0.2	0.1	0.1	0.2	0.1	0.2	0.1	0.1	NIL	NIL	0.1	0.1	0.3	0.2	NIL	NIL

TOTAL 3.6

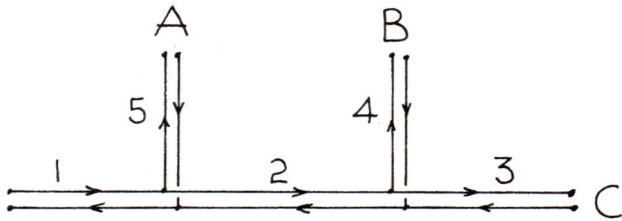

Fig. 3.12.

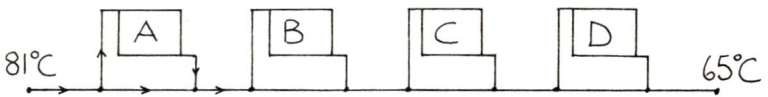

Fig. 3.13.

Heat load on 17 and 18 = 7.2 kW

$$\therefore \qquad \dot{m}_{17,18} = \frac{7.2}{4.2 \times 24.45} = 0.07 \text{ kg/s}$$

and $\qquad \dot{m}_{3,4} = 0.215 - 0.07 = 0.145 \text{ kg/s}$

The remaining calculations are similar to those above, and may be carried out as shown in the following tabulation.

Pipe No.	$\dot{m}$	Emission	Δt pipe	Initial temp.	Final temp.	Δt branch	Mean temp. at radiator
	kg/s	kW	°C	°C	°C	°C	°C
1	0.215	0.3	0.33	85	84.67 ⎫		
2	0.215	0.2	0.22	60	60.22 ⎭	24.45	
3	0.145	0.5	0.82	84.67	83.85 ⎫		
4	0.145	0.4	0.66	60.22	60.88 ⎭	22.97	
5	0.047	0.3	1.52	83.85	82.33		⎫ Rad.A:$\frac{1}{2}$(82.33+
6	0.047	0.2	1.01	60.88	61.89		⎭ 61.89)=72.1°C
7	0.098	0.1	0.24	83.85	83.61 ⎫		Rad.B:$\frac{1}{2}$(83.61+
8	0.098	0.1	0.24	60.88	61.12 ⎭	22.49	⎭ 61.12)=72.4°C
9	0.072	0.2	0.66	83.61	82.95 ⎫		
10	0.072	0.1	0.33	61.12	61.45 ⎭	21.50	
11	0.037	0.2	1.3	82.95	81.65		⎫ Rad.C:$\frac{1}{2}$(81.65+
12	0.037	0.1	0.65	61.45	62.10		⎭ 62.1) = 71.9°C
13	0.035	0.1	0.68	82.95	82.27		⎫ Rad.D:$\frac{1}{2}$(82.27+
14	0.035	0.1	0.68	61.45	62.13		⎭ 62.13)=72.2°C
15	0.026	NIL	–	83.61	83.61		⎫ Rad.B:
16	0.026	NIL	–	61.12	61.12		⎭ as above
17	0.07	0.1	0.34	84.67	84.33 ⎫		
18	0.07	0.1	0.34	60.22	60.56 ⎭	23.77	
19	0.03	0.3	2.4	84.33	81.93		⎫ Rad.F:$\frac{1}{2}$(81.93+
20	0.03	0.2	1.6	60.56	62.16		⎭ 62.16)=72°C
21	0.04	NIL	–	84.33	84.33		⎫ Rad.E:$\frac{1}{2}$(84.33+
22	0.04	NIL	–	60.56	60.56		⎭ 60.56)=72.4°C

It should be noted that the mean temperature of each radiator is approximately 72°C.

Problems

1. A two-pipe hot-water heating circuit is shown diagrammatically in Fig. 3.12. Using the data listed below, determine:

(a) the water flow rate in each pipe;

and   (b) the flow and return water temperature at each unit.

Data:

Unit		A	B	C
Emission, W		11 700	8 800	5 800

Pipe section		1	2	3	4	5
Emission	Flow main, W	1 900	1 000	900	400	150
	Return main, W	800	500	300	200	150

Specific heat capacity of water = 4.2 kJ/kg°C

Ans.: (*a*) 0.35, 0.21, 0.09, 0.12 and 0.14 kg/s
(*b*) A: 80.8 and 60.6°C, B: 80 and 61°C, C: 79.6 and 61.1°C.

2. A one-pipe hot-water system has four equal output radiators connected to it as shown in the sketch, Fig. 3.13. Determine the mean temperature of each radiator if the temperature drop across each radiator is 3°C above the minimum value.

Ans.: A: 77.5°C, B: 73.5°C, C: 69.5°C and D: 65.5°C.

N.B. Further examples and problems on hot-water heating systems will be found in Chapter 4.

4: Hot-Water Pipe Sizing

Introduction

Pipe sizing consists essentially of selecting the correct pipe diameter for the water flow rate found from the required heat emission and the permissible temperature drop. Since the total heat emission from any hot-water heating system is not known until the pipe sizes have been fixed, it becomes necessary, at the outset of the pipe-sizing procedure, to make a preliminary estimate of the water flow rate.

The water circulation may be either natural or forced, depending upon the size and use of the system. With natural circulation the pressure drop may be determined, as shown in Example 3.1.

With forced circulation it is usual to adopt a so-called "conventional" pressure drop when carrying out a preliminary pipe sizing of the index circuit. This pressure drop is tacitly assumed to be constant for the entire index circuit, regardless of changes in load that occur after branch connections. A few years ago a pressure drop of $100 - 150 \frac{N}{m^2}\Big/m$ of equivalent circuit length was usually adopted; this is now considered to be rather low, and a pressure drop within the range $150 - 300 \frac{N}{m^2}\Big/m$ is more common.

While these values are to some extent influenced by the type and range of pump available, they are often fixed quite arbitrarily, presumably based on the designer's experience. It would be fortuitous if the value of the pressure drop selected in this way was the correct one to give an economically sized pipe. The correct pressure drop is the one which results in minimum annual cost of owning and operating a pipe network. For a given flow rate a high pressure drop yields a smaller diameter, a lower capital cost and hence repayment, a lower heat emission from the pipe but a greater water pumping cost than a low-pressure drop. There is therefore an economic pressure drop, and hence diameter and velocity, for a given flow rate. The total annual cost will be a minimum when

$$\frac{d(A + B + C)}{dp} = 0 \tag{4.1}$$

where

A = annual cost of pumping. Found from: water horse-power, motor
and pump efficiency, electricity charges and running time.
B = annual repayment on capital expenditure. Found from: capital cost
of pipe line including supply, erection, thermal insulation and
establishment charges and profits, interest and depreciation
charges.
C = annual fuel cost due to waste heat emission from pipe. Found from:
running time, load factor, plant efficiency, calorific value of fuel
and fuel costs.
p = pressure drop.

If a fluid-flow formula of the exponential type is used the diameter in
items B and C may be expressed in terms of the flow rate and the pressure
drop.

Determining for each flow rate the economic pressure drop from
Eq (4.1) involves difficult and tedious calculations, particularly since many
of the factors involved are average values and depend to a great extent upon
the method of costing used by individual heating firms. For the smaller-
sized installation the time spent on detailed calculations may not be justi-
fied, but for large installations an economic solution is highly desirable. It
is possible that the time spent on calculations may be considerably
reduced by using electronic digital computers. Some work has been done
on this at the National College for Heating, Ventilating, Refrigeration and
Fan Engineering, and although a suitable computer programme has already
been drawn up and used to solve two problems, it is not yet clear to what
extent automatic computation will prove worthwhile.

Assuming average values, trial solutions of Eq (4.1) indicate that the
water velocity would be of the order of 1 to 2 m/s for water flow rates
between 0.01 and 100 kg/s.

Sometimes a total pressure is known at the outset of the pipe sizing, as
for example, when using a particular pump or extending an existing system.
In such cases the pressure drop will depend upon the circuit length, which,
for preliminary purposes, should include an allowance for local resistances.

Assuming that a suitable temperature difference between the flow and
return may be fixed, a convenient method of carrying out a trial-and-error
solution is outlined in the following procedure. The method consists of:
(*a*) preliminary calculations based on a number of assumptions, followed by
(*b*) a critical analysis to check the validity of the assumptions and to con-
firm the solution. The early success of this trial-and-error method depends
upon how nearly the designer can estimate the ratio of pipe emission to
heat load required by the units and upon realistic allowances for local
resistances.

(*a*) Preliminary calculations:

   1. For each pipe, tabulate or show on the drawings the following:

    (i) kW for the units to be served.
   (ii) kW for the pipes (usually assumed to be 10 to 30 per cent of the
       units emission depending upon whether the pipes are insulated
       or bare).
  (iii) kW for units plus pipes, i.e. (i) + (ii).
  (iv) kg/s flow rate, from (iii) divided by the selected temperature
      drop for the system and the specific heat capacity.

   2. *Pressure Drop or Water Velocity.* Adopt a pressure drop of, say,

$150-300 \dfrac{\text{N}}{\text{m}^2}\Big/\text{m}$ or a water velocity of $1-2$ m/s. Alternatively, if the

total pressure is known, determine the pressure drop by dividing the total
pressure by the total equivalent length of the index circuit. For preliminary purposes, allow an extra $10-50$ per cent on the actual length to
allow for local resistances.

   3. *Pipe Diameters.* These may now be selected from pipe-sizing tables
or charts, preferably from the current edition of the *I.H.V.E. Guide to
Current Practice.*

(*b*) Critical analysis:

   1. Estimate heat emission from the adopted diameters, and hence
obtain the total heat emission from all pipes. This may now be added
to the net heat requirement of all the units to obtain a revised estimate
of the total heat to be sent out from the source.

   2. Apportion mains emission to branches and obtain revised loads for
each pipe in the network.

   3. Using revised loads and selected diameters, find, from pipe-sizing
tables or charts, the pressure drop per unit length of pipe.

   4. Refer to tables of local resistances and obtain the total equivalent
length of each pipe.

   5. Calculate the total pressure drop for each pipe of the index circuit,
and hence the required pump head. Compare this with initial pressure, if
previously assumed. See (*a*) 2.

   6. Determine the pressure drop across each branch and check branch
pipe sizes.

While the above procedure is fairly straightforward, it has disadvantages:

   (*a*) the solution depends upon estimates, and a first-time balance is
unlikely;

   (*b*) branch circuits are difficult to analyse and balance;

   (*c*) no checks for economical pipe sizing.

The analysis takes a comparatively long time to complete and involves tedious and uninteresting calculations. For this reason many designers complete the preliminary calculations only and rely on site adjustment of lockshield values to make up for any deficiencies of balance at branches. Without controls many systems would fail to behave in the manner planned and laid down by the initial design criteria.

It is convenient to assume a constant temperature drop across each branch immediately after a junction, that is, the temperature of the water in each return pipe immediately before the junction is the same. When one of the branches is considerably longer than the other it may be advantageous to keep its diameter to a minimum. This may be achieved by arranging a greater temperature drop across the longer branch, and hence a lower flow rate. The return temperature after the junction would then be the resultant mixture temperature of water from the two returns. Extra calculations are necessary if a constant temperature drop for each branch is not assumed, but the reduced pipe and installation costs may make them worthwhile.

A heating circuit consists essentially of straight pipe and local resistances such as valves, bends, tees and general changes in shape and direction. The following outlines the basic principles involved when determining the pressure loss in such systems and is particularly useful in the solution of simple networks as in Example 4.9.

Consider first the straight pipe:

Loss of head (h) of fluid flowing for a circular pipe running full is given by the rational equation:

$$h = \frac{4flu^2}{2gd} \qquad (4.2)$$

where f = coefficient of friction
l = length of pipe
u = velocity of water flow
d = internal diameter of pipe
g = acceleration due to gravity.

Loss of pressure $\Delta p = h\rho g$ \qquad\qquad (4.3)

therefore $h = \dfrac{\Delta p}{\rho g} = \dfrac{4flu^2}{2gd}$ \qquad\qquad (4.4)

where ρ = density of water

Since $\dot{m} = \dfrac{\rho \pi d^2 u}{4}$

$$u = \frac{4\dot{m}}{\rho \pi d^2}$$

Substituting for u^2 in Eq (4.4) we have:

$$\frac{\Delta p}{\rho g} = \frac{4fl}{2gd} \left(\frac{4\dot{m}}{\rho \pi d^2}\right)^2 \tag{4.5}$$

If 'f' is assumed to be constant (the coefficient of friction is in fact a variable dependent upon the physical characteristics of the fluid flowing, the Reynolds Number and the roughness of the pipe surface relative to the internal diameter. The error involved by assuming 'f' to be constant is, however, small) then Eq (4.5) becomes

$$\frac{\Delta p}{l} = C\frac{\dot{m}^2}{d^5} \tag{4.6}$$

where C = a constant

The loss of pressure in the local resistances is generally expressed in terms of the velocity pressure, i.e.

$$\Delta p = K\tfrac{1}{2}\rho u^2 \tag{4.7}$$

where K = velocity pressure factor.

Consider a straight pipe having a length 'l' from 'a' to 'b' and containing a valve, then if $\Delta p/l$ = the pressure loss per unit length of straight pipe, from Table 4.1, then

$$\Delta p_{a \to b} = \left(\frac{\Delta p}{l}\right)l + K\tfrac{1}{2}\rho u^2$$

put

$$\tfrac{1}{2}\rho u^2 = \left(\frac{\Delta p}{l}\right)l_e$$

where l_e = length of straight pipe having a pressure loss equal to one velocity pressure; i.e. K = 1.0.

Then

$$\Delta p_{a \to b} = \frac{\Delta p}{l}\left[l + K \times l_e\right] \tag{4.8}$$

Values of $\Delta p/l$, l_e and K are given in Tables 4.1, 4.2, 4.3 and 4.4 which are based on the data given in the National College for Heating, Ventilating, Refrigeration and Fan Engineering, Technical Memorandum No. 7* and in the current edition of the IHVE Guide to Current Practise. These tables will be used in some of the examples that follow.

* Fluid flow in S.I. units, A. F. Armor, M. J. Farrell and B. G. Lawrence, Oct. 1968.

TABLE 4.1

Flow of Water at 75°C in Steel Tubes
(For heavy grade tubes to B.S. 1387: 1967)

Diameter, mm

$\Delta p/l$	15 $\dot m$	15 l_e	20 $\dot m$	20 l_e	25 $\dot m$	25 l_e	32 $\dot m$	32 l_e	40 $\dot m$	40 l_e	50 $\dot m$	50 l_e	65 $\dot m$	65 l_e	80 $\dot m$	80 l_e	Approx. Velocity u
40	0.031	0.4	0.072	0.6	0.133	0.9	0.292	1.3	0.447	1.6	0.853	2.2	1.739	3.1			0.5
50	0.035	0.4	0.081	0.6	0.150	0.9	0.330	1.3	0.504	1.6	0.962	2.2	1.958	3.2			
67.5	0.041	0.4	0.096	0.7	0.177	0.9	0.388	1.3	0.592	1.65	1.128	2.3	2.295	3.2			
75	0.043	0.4	0.101	0.7	0.188	0.9	0.410	1.34	0.620	1.7	1.190	2.3	2.427	3.3			
77.5	0.044	0.4	0.103	0.7	0.191	0.9	0.418	1.34	0.638	1.7	1.214	2.3	2.469	3.3			
80	0.045	0.4	0.105	0.7	0.194	0.9	0.425	1.35	0.649	1.7	1.248	2.3	2.510	3.3			
82.5	0.046	0.4	0.107	0.7	0.197	0.9	0.432	1.35	0.659	1.7	1.260	2.3	2.550	3.3			
95	0.049	0.4	0.115	0.7	0.231	0.9	0.466	1.36	0.711	1.7	1.352	2.3	2.748	3.3			
120	0.056	0.4	0.131	0.7	0.242	0.9	0.527	1.4	0.805	1.7	1.530	2.4	3.106	3.3	4.819	4.1	1.0
140	0.061	0.5	0.142	0.7	0.260	0.9	0.572	1.4	0.873	1.7	1.659	2.4	3.367	3.4	5.222	4.2	
160	0.065	0.5	0.152	0.7	0.282	1.0	0.614	1.4	0.937	1.7	1.880	2.4	3.610	3.4	5.598	4.2	
180	0.070	0.5	0.162	0.7	0.300	1.0	0.654	1.4	0.997	1.8	1.890	2.4	3.838	3.4	5.950	4.2	
200	0.074	0.5	0.172	0.7	0.317	1.0	0.691	1.4	1.053	1.8	1.990	2.4	4.054	3.4	6.285	4.2	
220	0.078	0.5	0.181	0.7	0.334	1.0	0.727	1.4	1.107	1.8	2.101	2.4	4.260	3.4	6.600	4.2	
240	0.081	0.5	0.189	0.7	0.349	1.0	0.761	1.4	1.159	1.8	2.199	2.4	4.460	3.4	6.910	4.2	
260	0.085	0.5	0.198	0.7	0.364	1.0	0.793	1.5	1.210	1.8	2.292	2.4	4.650	3.4	7.200	4.2	1.5
280	0.088	0.5	0.206	0.7	0.379	1.0	0.825	1.5	1.260	1.8	2.382	2.4	4.830	3.4	7.480	4.3	
300	0.092	0.5	0.213	0.7	0.393	1.0	0.855	1.5	1.302	1.8	2.469	2.5	5.000	3.5	7.750	4.3	
400	0.107	0.5	0.248	0.7	0.457	1.0	0.994	1.5	1.513	1.8	2.870	2.5	5.800	3.5	8.990	4.3	
420	0.110	0.5	0.255	0.7	0.469	1.0	1.019	1.5	1.551	1.8	2.940	2.5	5.950	3.5	9.220	4.3	
460	0.115	0.5	0.267	0.7	0.492	1.0	1.069	1.5	1.626	1.8	3.080	2.5	6.240	3.5	9.660	4.3	
480	0.118	0.5	0.273	0.75	0.503	1.0	1.093	1.5	1.663	1.8	3.150	2.5	6.370	3.5	9.870	4.3	
500	0.120	0.5	0.279	0.75	0.514	1.0	1.116	1.5	1.698	1.8	3.220	2.5	6.510	3.5	10.100	4.3	

NOTE: $\Delta p/l$ = pressure loss per unit length $\dfrac{N}{m^2}/m$

$\dot m$ = mass flow rate kg/s

l_e = equivalent length of pipe in metres for K = 1.0

u = water velocity m/s

K = velocity pressure factor, see Table 4.4

TABLE 4.2

Correction Factors for Flow of Water at 75°C in
Medium Grade Steel Pipes to B.S. 1387: 1967

Correction factor to be applied to values given in Table 4.1	Diameter, mm						
	15	25	32	40	50	65	80
$\phi_{\Delta p}$	0.7	0.75	0.79	0.83	0.85	0.87	0.91
ϕ_{l_e}	1.25	1.1	1.07	1.05	1.04	1.03	1.02

TABLE 4.3

Correction Factor for Flow of Water at 150°C in
Heavy Grade Steel Pipes to
B.S. 1387: 1967

$\Delta p/l$ $\frac{N}{m^2}$ per metre length from Table 4.1	Diameter, mm					
	25		50		100	
	$\phi_{\Delta p}$	ϕ_{l_e}	$\phi_{\Delta p}$	ϕ_{l_e}	$\phi_{\Delta p}$	ϕ_{l_e}
50	0.92	1.2	0.95	1.1	0.98	1.1
100	0.94	1.2	0.97	1.1	0.99	1.1
200	0.96	1.1	0.98	1.1	1.0	1.1
500	0.98	1.1	0.99	1.1	1.02	1.0

TABLE 4.4

Velocity Pressure Factors (K)

Tees and Junctions (based on velocity pressure of combined flow).

(a) To or from 90° Branch: 0.5 + Bend factor + Reduction or enlargement factor.
(b) To or from run, i.e. straight through tee: 0.2 + Reduction or enlargement factor.

Reductions and Enlargements (based on velocity pressure in smaller pipe).

(a) Reductions:

Diam. ratio 3 : 2 K = 0.3
Diam. ratio 4 : 1 K = 0.5

(b) Enlargements:

Diam. ratio 3 : 2 K = 0.4
Diam. ratio 4 : 1 K = 1.0

Bends.

(a) Malleable Cast Iron:

Diam. 15−25 mm K = 0.7
Diam. 32−50 mm K = 0.5
Diam. 65−90 mm K = 0.4

(b) Welded Wrought Iron:

Diam. 32−50 mm K = 0.3

Valves.

(a) Gate valve   K = 0.2
(b) Globe valve   K = 5.0
(c) Pillar tap   K = 10.0

Radiators.   K = 5.0

Example 4.1. Part of a hot-water heating system is shown in Fig. 4.1. Using the data listed below and assuming that Medium Grade tubing is used with Malleable Cast Iron fittings, determine:

    (*a*) the pressure drop from A to B;
    (*b*) the ratio of fittings length to actual length.

Temperature of water: 75°C.

Pipe No.	1	2	3	4	5	6
Length l, m	5	6	15	20	7	7
Flow rate ($\dot{m}$), kg/s .	2.55	1.89	1.26	0.761	0.242	0.065
Diameter (d), mm ..	65	50	40	32	25	15
K factor	1.7	0.5	1.8	2.5	1.6	8.2

Since the main pipe sizing table is for Heavy Grade tubing, it will be necessary to use the correction factors given in Table 4.2. The pressure drop along each pipe, and hence the total pressure drop, is found as shown in the following tabulation:

(*a*)

		1	2	3	4	5	6
1	Pipe number	1	2	3	4	5	6
2	Pipe diameter, mm	65	50	40	32	25	15
3	Water flow rate, kg/s	2.55	1.89	1.26	0.761	0.242	0.065
4	$\Delta p/l, \frac{N}{m^2}/m$, Table 4.1	82.5	180	280	240	120	160
5	l_e, m, Table 4.1	3.3	2.4	1.8	1.4	0.9	0.5
6	$\phi_{\Delta p}$, Table 4.2	0.87	0.85	0.83	0.79	0.75	0.7
7	ϕ_{l_e}, Table 4.2....................	1.03	1.04	1.05	1.07	1.1	1.25
8	l_e medium grade, m (5 × 7)	3.4	2.5	1.89	1.5	0.99	0.63
9	$\Delta p/l$, medium grade, $\frac{N}{m^2}/m$, (4 × 6) ..	71.8	153	232	190	90	112
10	K	1.7	0.5	1.8	2.5	1.6	8.2
11	K × l_e, m, (8 × 10)	5.78	1.25	3.4	3.75	1.58	5.17
12	l, m	5	6	15	20	7	7
13	Total equivalent length, m, (11 + 12) .	10.78	7.25	18.4	23.75	8.58	12.17
14	$\Delta p, \frac{N}{m^2}$, (9 × 13)	744	1 109	4 269	4 513	772	1 363

Total pressure loss from A to B = sum of line 14

$$= 12.77 \text{ kN/m}^2$$

(*b*)   Total length of pipe (line 12)   = 60 m
            Total length of fittings (line 11) = 21 m

            Total equivalent length   = 81 m

From which the total length of the fittings is found to be:

$$\frac{21}{60} \times 100 = 35 \text{ per cent}$$

of the total length of the pipe.

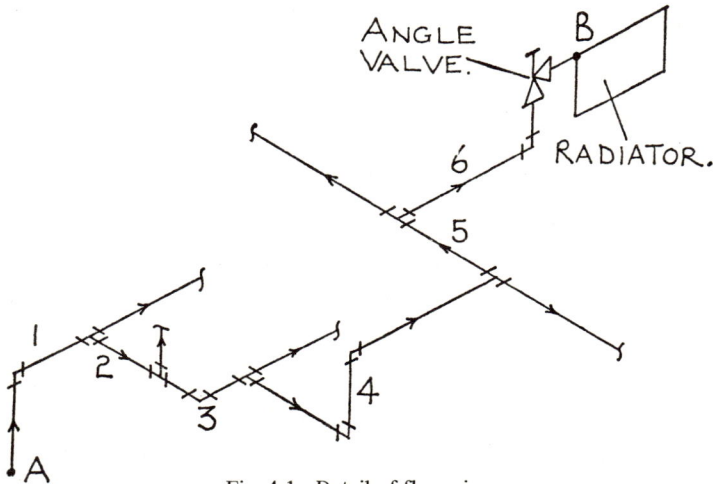

Fig. 4.1. Detail of flow pipes.

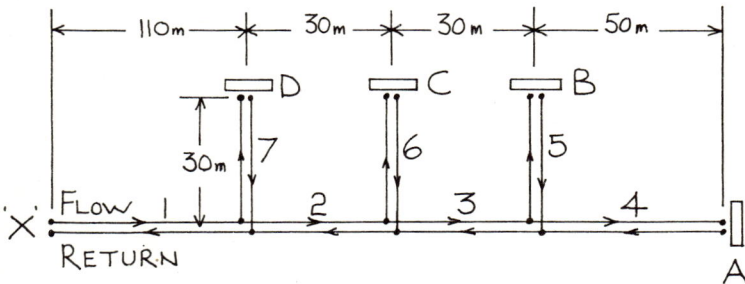

Fig. 4.2 Hot-water heating system.

Example 4.2. A forced-circulation hot-water heating system is shown diagrammatically in Fig. 4.2. If the pressure drop across the system at "X" is 22 kN/m², and assuming that the lengths given include an allowance for local resistances, determine:

(*a*) the average pressure drop for (i) the index circuit to unit A and (ii) the branch circuit to unit C;

(*b*) the diameter of the pipes in sections 1 to 4 and 6.

Data:

Mean temperature of water = 75°C.
Tubing: Heavy Grade.

Pipe No	1	2	3	4	6
Flow rate, kg/s ..	1.51	1.22	0.98	0.6	0.24

(*a*) Total length of index circuit to unit A = 440 m (flow + return)

Then average pressure drop $= \dfrac{22 \times 10^3}{440} = 50 \ \dfrac{\text{N}}{\text{m}^2}\Big/\text{m}$

Since the various circuits are arranged in parallel

$$\Delta p_6 = \Delta p_3 + \Delta p_4$$

where Δp = pressure drop

and the subscript refers to the pipe section.

Length of section 6   =   60 m
Length of section 3 + 4 = 160 m

Then average pressure drop for branch 6 to unit C will be:

$$\dfrac{160 \times 50}{60} = 133 \ \dfrac{\text{N}}{\text{m}^2}\Big/\text{m}$$

Alternatively,

$$\Delta p_6 = \Sigma \Delta p - (\Delta p_1 + \Delta p_2)$$
Length of section 1 + 2 = 280 m

Then

$$\Delta p_6 = 22 \times 10^3 - 280 \times 50 = 8\,000 \ \dfrac{\text{N}}{\text{m}^2}$$

and average pressure drop will be

$$\dfrac{8\,000}{60} = 133 \ \dfrac{\text{N}}{\text{m}^2}\Big/\text{m}$$

as before.

(*b*) Referring to the pipe sizing, Table 4.1, the diameters required are as follows:

Pipe No.	Flow rate, kg/s	Pressure drop, $\dfrac{N}{m^2}\big/m$	Diam., mm
1	1.51	50	65
2	1.22	50	65
3	0.98	50	50
4	0.60	50	50
6	0.24	133	25

Note the difficulty of selecting diameters from the commercial range to suit the given flow rates and pressure drops. In each case the nearest commercial diameter has been selected, and in this case the system is oversized. The total pressure drop for the index circuit will therefore be less than the 22 kN/m² available. The surplus pressure may in practice be absorbed by valve restriction.

Example 4.3. Some details of the index circuit of a hot-water heating system are given below. The mean temperature of the water is 150°C, and Heavy Grade tubing is used throughout.

(*a*) Determine the total resistance of the circuit.

(*b*) If the pump is installed in the flow pipe and the flow water temperature is 170°C, what will be its actual duty?

Data:

Pipe section No.	1	2	3	4	5	6
Pipe section length, flow + return, m′	150	60	60	250	100	120
Pipe section flow rate, kg/s	5.95	2.51	1.89	1.24	0.873	0.627
Pipe section diameter, mm	80	65	50	50	40	40
Velocity pressure factor, flow + return ..	8	6	7	3	5	10

Specific volume of water at 170°C = 0.1114×10^{-2} m^3/kg

(*a*) Using the data from Table 4.1 and Table 4.3, the pressure drop will be as follows:

		1	2	3	4	5	6
1	Pipe section number	1	2	3	4	5	6
2	″ ″ diameter mm ..	80	65	50	50	40	40
3	$\Delta p/l$, $\frac{N}{m^2}$/m, Table 4.1	180	80	180	80	140	75
4	l_e, m, Table 4.1	4.2	3.3	2.4	2.3	1.7	1.7
5	$\phi_{\Delta p}$, Table 4.3 (approx.) ..	0.98	0.98	0.98	0.97	0.96	0.95
6	ϕ_{l_e}, Table 4.3 (approx.) ..	1.1	1.1	1.1	1.1	1.1	1.2
7	(3×5), $\frac{N}{m^2}$/m	177	78	177	78	135	71
8	(4×6), m	4.6	3.6	2.6	2.5	1.9	2.0
9	K	8	6	7	3	5	10
10	(9×8), m	36.8	21.6	18.2	7.5	9.5	20
11	l, m	150	60	60	250	100	120
12	$(10 + 11)$, m	186.8	81.6	78.2	257.5	109.5	140
13	(12×7), $\frac{N}{m^2}$	23 064	6 365	13 841	20 085	14 783	9 940

Total pressure loss = sum of line 13 = 88 kN/m^2 approx.

(*b*) Pump duty = $5.95 \times 0.1114 \times 10^{-2} \times 10^3$ = 6.63 l/s approx.

Example 4.4. It is required to circulate 2.5 kg/s of water at 75°C through each of the branch circuits A, B and C of the heating system shown diagrammatically in Fig. 4.3. Determine by how much the resistance of the branch circuits A and B must be increased so that the whole system is in balance. Assume that Heavy Grade tubing is used and that the lengths given include an allowance for local resistances.

Pipe No.	1	2	3	4	5	6
Diameter, mm	80	80	65	80	65	65
Length, m	60	30	90	30	45	45

Analyse the system as follows, using the data given in Table 4.1:

Pipe No.	Flow rate $(\dot{m})$, kg/s	Diam., mm	Total equivalent length, m	Pressure drop, $\frac{N}{m^2}/m$	Pressure drop, $\frac{N}{m^2}$	
1	7.5	80	60	280	14 800	
2	5.0	80	30	130	3 900	29 800
3	2.5	65	90	80	7 200	
4	5.0	80	30	130	3 900	
5	2.5	65	45	80	3 600	
6	2.5	65	45	80	3 600	

Index circuit = Pipes 1, 2, 3 and 4 (because $\Delta p_3 > \Delta p_5$)
Pressure drop = 29 800 N/m²
Pressure available for sizing Branch A (pipe 6)

$$= 29\,800 - \Delta p_1$$
$$= 29\,800 - 14\,800 = 15\,000 \text{ N/m}^2$$

Pressure absorbed by Branch A (pipe 6) when passing the required 2.5 kg/s = 3 600 N/m².

∴ Increase in resistance required $= 15\,000 - 3\,600$
$$= 11\,400 \text{ N/m}^2$$

Pressure available for sizing Branch B (pipe 5)

$$= \text{pressure drop for Branch C (pipe 3)}$$
$$= 7\,200 \text{ N/m}^2$$

Pressure absorbed by Branch B (pipe 5) when passing the required 2.5 kg/s = 3 600 N/m².

∴ Increase in resistance required $= 7\,200 - 3\,600$
$$= 3\,600 \text{ N/m}^2$$

Example 4.5. The index circuit of a hot-water heating system designed to operate at a mean water temperature of 75°C is detailed in the table below.

(*a*) If Heavy Grade tubing is used, determine the total resistance of the circuit.

(*b*) What will be the diameter of the branch taken from the index circuit at the end of section 3 if its length, including an allowance for local resistances, is 50 m. By how much should the resistance of this branch be increased to equal the pressure available?

Section No.	1	2	3	4	5
Equivalent length (flow + return), m	25	40	30	35	20
Diameter, mm	50	40	25	20	15
Flow rate, kg/s	1.26	0.638	0.262	0.115	0.056

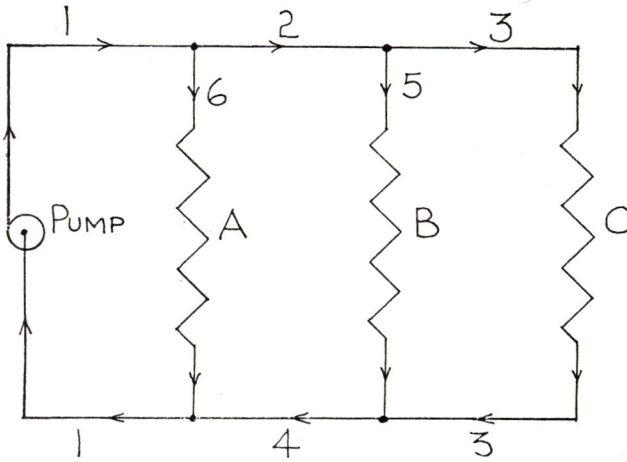

Fig. 4.3. Diagram of hot-water heating system

(*a*) From Table 4.1:

Section No.	Pressure drop, $\dfrac{N}{m^2}/m$	Equivalent length, m	Pressure drop, $\dfrac{N}{m^2}$
1	82.5	25	2 063
2	77.5	40	3 100
3	140	30	4 200
4	95	35	3 325
5	120	20	2 400
		Total resistance	15 088 N/m² $= 15.088 \text{ kN/m}^2$

(*b*) Pressure available at end of Section 3

$$= 15\,088 - \Delta p_{1\text{-}3}$$
$$\text{or,} = \Delta p_4 + \Delta p_5$$
$$= 5\,725 \text{ N/m}^2$$

Pressure drop for branch $= \dfrac{5\,725}{50} = 114.5 \dfrac{N}{m^2}/m$

Flow rate for branch $= \dot{m}_3 - \dot{m}_4$
$$= 0.262 - 0.115 = 0.147 \text{ kg/s}$$

Then from Table 4.1, nearest diameter = 25 mm

Actual pressure drop for 25 mm pipe passing 0.147 kg/s = $47.5 \dfrac{N}{m^2}/m$

and total pressure drop = 50 × 47.5 = 2 375 N/m².

Additional resistance required

$$= 5\,725 - 2\,375 = 3\,350 \text{ N/m}^2 = 3.350 \text{ kN/m}^2$$

Example 4.6. Carry out a preliminary pipe sizing of the hot-water system shown in Fig. 4.4. Use the following data and assumptions:

Heat emission from each unit (A to E)	12 kW
Temperature drop across system	15°C
Mean temperature of water	75°C
Heat emission from pipes as a percentage of emission from units	30 per cent
Specific heat capacity of water	4.2 kJ/kg/°C

Pressure drop to be within the range of $150 - 500 \dfrac{\text{N}}{\text{m}^2}\Big/\text{m}$.

Preliminary water flow rates ($\dot{m}$), kg/s:

$$\dot{m}_1 = \frac{1.3 \times 5 \times 12}{4.2 \times 15} = 1.24 \text{ kg/s}$$

$$\dot{m}_2 = \frac{1.3 \times 4 \times 12}{4.2 \times 15} = 0.99 \text{ kg/s}$$

$$\dot{m}_3 = \frac{1.3 \times 3 \times 12}{4.2 \times 15} = 0.744 \text{ kg/s}$$

$$\dot{m}_4 = \frac{1.3 \times 2 \times 12}{4.2 \times 15} = 0.495 \text{ kg/s}$$

$$\dot{m}_{5-9} = \frac{1.3 \times 12}{4.2 \times 15} = 0.248 \text{ kg/s}$$

Then from Table 4.1 the pipe diameters are:

Section No.	1	2	3	4	5–9
Diameter, mm	40	40	32	25	20

Example 4.7. Determine the total resistance of the index circuit of the system shown in Fig. 4.4. Use the diameters found in the previous question and the following data:

Tubing: Medium Grade.

Section No.		1	2	3–9
Length, Flow + Return, m		30	30	20
Velocity pressure factor (K)		2	3	2
Pipe diameter, mm	40	32	25	20
Heat emission, W/m	135	115	96	75

By inspection the index circuit consists of sections 1–5.

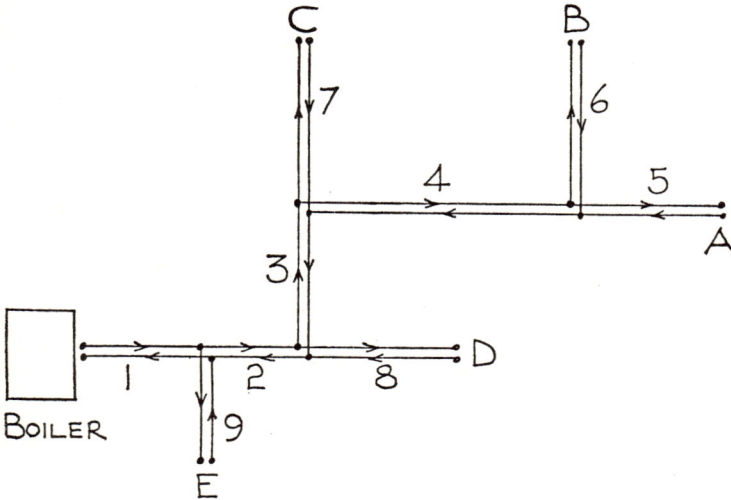

Fig. 4.4. Two-pipe system.

Then:

Revised heat load:

Section	Length, m	W/m	W
1	30	135	4 050
2	30	135	4 050
3	20	115	2 300
4	20	96	1 920
5	20	75	1 500
6	20	75	1 500
7	20	75	1 500
8	20	75	1 500
9	20	75	1 500

Total heat emission from the pipes = 19 820
" " " " " units = 60 000

Total heat emission = 79 820 W, i.e. 79.82 kW

Revised flow rates ($\dot{m}$), kg/s:

$$\dot{m}_1 = \frac{79.82}{4.2 \times 15} = 1.27 \text{ kg/s}$$

$$\Delta t_1 = \frac{4.05}{4.2 \times 1.27} = 0.76°\text{C}$$

Temperature drop across branch at end of section 1

$$= 15 - 0.76 = 14.24°\text{C}$$

Total heat load on Section 2 = 79.82 − 12 − 4.05 = 63.77 kW

$$\dot{m}_2 = \frac{63.77}{4.2 \times 14.24} = 1.07 \text{ kg/s}$$

$$\Delta t_2 = \frac{4.05}{4.2 \times 1.07} = 0.9°\text{C}$$

Temperature drop across branch at end of Section 2

$$= 14.24 - 0.9 = 13.34°\text{C}$$

Total heat load on section 3 = 44.72 kW

$$\dot{m}_3 = \frac{44.72}{4.2 \times 13.34} = 0.8 \text{ kg/s}$$

$$\Delta t_3 = \frac{2.3}{4.2 \times 0.8} = 0.69°\text{C}$$

Temperature drop across branch at end of section 3

$$= 13.34 - 0.69 = 12.65°\text{C}$$

Total heat load on section 4 = 27 kW

$$\dot{m}_4 = \frac{27}{4.2 \times 12.65} = 0.51 \text{ kg/s}$$

Then, since the heat load on section 5 is the same as the heat load on section 6, the flow rate $\dot{m}_4$ will be divided equally between sections 5 and 6.

$$\dot{m}_5 = \frac{0.51}{2} = 0.255 \text{ kg/s}$$

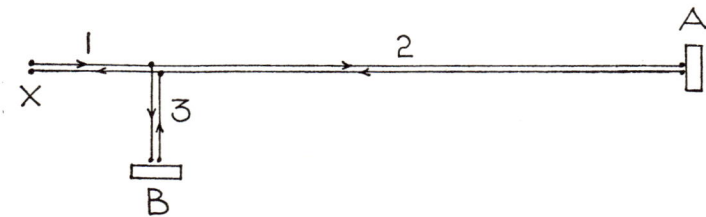

Fig. 4.5. A comparison between long and short branches.

The total resistance of sections 1—5 may now be found as shown in the following tabulation. Use the data given in Tables 4.1 and 4.2.

		1	2	3	4	5
1	Section number	1	2	3	4	5
2	Diameter, mm	40	40	32	25	20
3	Flow rate, kg/s	1.27	1.07	0.8	0.51	0.255
4	$\Delta p/l$, $\frac{N}{m^2}$/m, Table 4.1	285	212	260	500	420
5	l_e, m, Table 4.1	1.8	1.8	1.5	1.0	0.7
6	$\phi_{\Delta p}$, Table 4.2	0.83	0.83	0.79	0.75	0.72
7	ϕ_{l_e}, Table 4.2	1.05	1.05	1.07	1.1	1.18
8	(4 × 6), $\frac{N}{m^2}$/m	237	176	205	375	302
9	(5 × 7), m	1.89	1.89	1.61	1.1	0.83
10	K	2	3	2	2	2
11	(9 × 10), m..............	3.78	5.77	3.22	2.2	1.66
12	l, m	30	30	20	20	20
13	(11 + 12), m	33.78	35.77	23.22	22.2	21.66
14	(8 × 13), $\frac{N}{m^2}$	8 000	6 300	4 550	8 325	6 550

Total resistance = sum of line 14 = 33.725 kN/m²

Example 4.8. Determine and compare the pipe diameters for the hot-water system shown in Fig. 4.5 if:

(*a*) the temperature drop across each branch is 12°C;

(*b*) the temperature drop across the long branch, section 2, is 24°C and across the short branch, section 3, is 12°C

Assume Heavy Grade tube and water at 75°C

Also, what will be the return water temperature of section 1 for conditions given in (*b*) if the flow water temperature is 80°C.

Data:

Specific heat capacity of water = 4.2 kJ/kg°C
Heat emission from each unit   = 30 kW
Pressure available at "X"   = 30 kN/m²

Section No.	1	2	3
Length, flow + return, m ...	70	350	70

(*a*)
$$\dot{m}_2 = \frac{30}{4.2 \times 12} = 0.595 \text{ kg/s}$$

$$\dot{m}_3 = \frac{30}{4.2 \times 12} = 0.595 \text{ kg/s}$$

$$\dot{m}_1 = \dot{m}_2 + \dot{m}_3 = 1.19 \text{ kg/s}$$

By inspection, the index circuit consists of sections 1 and 2 to unit A.

$$\text{Pressure drop} = \frac{30 \times 10^3}{70 + 350} = 71.5 \frac{N}{m^2}\bigg/m$$

Then from the pipe-sizing tables:

$$d_1 = 50 \text{ mm (Pressure drop} = 75 \frac{N}{m^2}\bigg/m)$$

$$d_2 = 40 \text{ mm (Pressure drop} = 67.5 \frac{N}{m^2}\bigg/m)$$

Then, for parallel circuits

$$\Delta p_3 = \Delta p_2$$

$$= 350 \times 67.5 = 23\,625 \frac{N}{m^2}$$

i.e. $\quad \dfrac{\Delta p}{l} = \dfrac{23\,625}{70} = 337.5 \dfrac{N}{m^2}\bigg/m)$

and

$$d_3 = 32 \text{ mm (nearest commercial size. Pressure drop}$$

$$= 150 \frac{N}{m^2}\bigg/m)$$

(b) $\quad \dot{m}_2 = \dfrac{30}{4.2 \times 24} = 0.298 \text{ kg/s}$

$$\dot{m}_3 = 0.595 \text{ as in } (a)$$

$$\dot{m}_1 = \dot{m}_2 + \dot{m}_3 = 0.893 \text{ kg/s}$$

$$\Delta p_{1,2} = 71.5 \frac{N}{m^2}\bigg/m \text{ as in } (a)$$

Then $\quad d_1 = 40 \text{ mm (Pressure drop } 147 \dfrac{N}{m^2}\bigg/m \text{ approx.)}$

$$d_2 = 32 \text{ mm (Pressure drop } 41.5 \frac{N}{m^2}\bigg/m \text{ approx.)}$$

$$\Delta p_3 = \Delta p_2$$

$$= 350 \times 41.5 = 14\,525 \frac{N}{m^2}$$

i.e. $\quad \dfrac{\Delta p}{l} = \dfrac{14\,525}{70} = 207.5 \dfrac{N}{m^2}\bigg/m$

and

$$d_3 = 32 \text{ mm (nearest commercial sizes. Pressure drop}$$

$$= 150 \frac{N}{m^2}\bigg/m)$$

Comparison of diameters:

Section No.	1	2	3
Case (*a*)	50 mm	40 mm	32 mm
(*b*)	40 mm	32 mm	32 mm

Some economy would result from adopting 24°C across the long branch.
Return water temperature (t_r) in section 1 for case (*b*), by method of mixtures:

$$\frac{0.298(80 - 24) + 0.595(80 - 12)}{0.298 + 0.595} = 64.3°C \text{ approx.}$$

Example 4.9. A hot-water heating system is shown diagrammatically in Fig. 4.6. Using the design details given below, calculate:

(*a*) The water flow rate in each section of the system.
(*b*) The flow and return water temperature at unit A.
(*c*) The total pressure loss in the index circuit if the pressure loss in section 3 is fixed at 285 N/m².
(*d*) The extra resistance required in the sub-circuits to balance the system.

Data:

Pipe section number		1	2	3	4	5
Pipe heat emission	flow main kW	1.9	1.0	0.9	0.4	0.15
	return main kW ...	0.8	0.5	0.3	0.2	0.15
Total equivalent length (flow + return), m		60	30	60	30	15
Diameter, mm		50	40	32	25	25

Unit	A	B	C
Heat emission, kW	11.7	8.8	5.8

Pressure loss across boiler	= 700 N/m²
Pressure loss across each unit	= 500 N/m²
Specific heat capacity of water	= 4.2 kJ/kg°C
Flow water temperature at boiler	= 82.2°C
Return water temperature at boiler	= 60.0°C

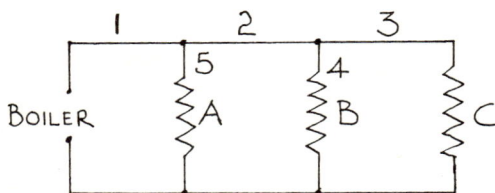

Fig. 4.6.

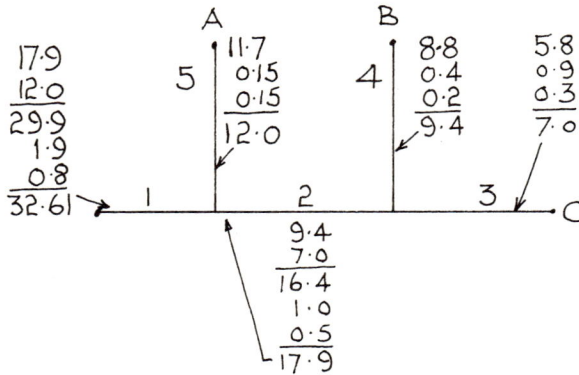

Fig. 4.7.

Solution (*a*)

Total heat emission = 26.3 kW (units) + 6.3 kW (pipes) = 32.6 kW

$$\therefore \qquad \dot{m}_1 = \frac{32.6}{4.2(82.2 - 60)} = 0.35 \text{ kg/s}$$

This may be proportional to the various sections of the system by simple ratio using the data given in Fig. 4.7 and as shown in the previous chapter.

$$\dot{m}_2 = \frac{17.9}{29.9} \times 0.35 = 0.21 \text{ kg/s}$$

$$\dot{m}_3 = \frac{7.0}{16.4} \times 0.21 = 0.09 \text{ kg/s}$$

$$\dot{m}_4 = \dot{m}_2 - \dot{m}_3 = 0.12 \text{ kg/s}$$

$$\dot{m}_5 = \dot{m}_1 - \dot{m}_2 = 0.14 \text{ kg/s}$$

(*b*) By calculating the temperature drop along the flow (F) and return (R) mains:

$$(t_F)_A = 82.2 - \frac{1.9}{0.35 \times 4.2} - \frac{0.15}{0.14 \times 4.2} = 80.7°C$$

$$(t_R)_A = 60 + \frac{0.8}{0.35 \times 4.2} + \frac{0.15}{0.14 \times 4.2} = 60.8°C$$

(*c*) Since the pressure loss in section 3 is fixed the pressure loss in the remaining sections may be found by applying Eq (4.6) as follows:

The pressure loss (Δp) per unit length (*l*) in section 3 will be:—

$$\left(\frac{\Delta p}{l}\right)_3 = \frac{285}{60} = 4.75 \frac{N}{m^2}\Big/m$$

Then

$$\left(\frac{\Delta p}{l}\right)_2 = 4.75\left(\frac{0.21}{0.09}\right)^2\left(\frac{32}{40}\right)^5 = 8.47 \frac{N}{m^2}\Big/m$$

$$\left(\frac{\Delta p}{l}\right)_1 = 4.75\left(\frac{0.35}{0.09}\right)^2\left(\frac{32}{50}\right)^5 = 7.71 \frac{N}{m^2}\Big/m$$

$$\left(\frac{\Delta p}{l}\right)_4 = 4.75\left(\frac{0.12}{0.09}\right)^2\left(\frac{32}{25}\right)^5 = 29.02 \frac{N}{m^2}\Big/m$$

$$\left(\frac{\Delta p}{l}\right)_5 = 4.75\left(\frac{0.14}{0.09}\right)^2\left(\frac{32}{25}\right)^5 = 39.5 \frac{N}{m^2}\Big/m$$

Thus

$\Delta p_1 = 60 \times 7.71 = 463$ N/m^2

$\Delta p_2 = 30 \times 8.47 = 254$ "

$\Delta p_3 =$ (given) $= 285$ "

$\Delta p_4 = 30 \times 29.02 = 871$ "

$\Delta p_5 = 15 \times 39.5 = 593$ "

By inspection the index circuit consists of sections 1, 2 and 4, the boiler and unit B; its resistance will be:$- 463 + 254 + 871 + 700 + 500 = 2\,788$ N/m^2.

(d) Since the sub-circuits are in parallel with the index circuit we have:

$$\Delta p_3 + (\Delta p_3)_{\text{balance}} + \Delta p_C = \Delta p_4 + \Delta p_B$$

and since $\Delta p_C = \Delta p_B$, $(\Delta p_3)_{\text{balance}} = \Delta p_4 - \Delta p_3 = 586$ N/m^2.

Similarly,

$$(\Delta p_5)_{\text{balance}} = \Delta p_2 + \Delta p_4 - \Delta p_5 = 532 \text{ N/m}^2$$

These extra resistances may be provided by valve regulation or by installing orifice plates.

Problems

1. Some details of a two-pipe hot-water heating system are given in Fig. 4.8 and the table below. At "X" the flow and return water temperatures are 82.2°C and 71.1°C respectively and the total pressure available is 44.79 kN/m^2. A, B, C and D are heating units. Determine:

(a) The approximate sizes of all pipes.

(b) The resistance of the index circuit. Assume, Heavy Grade tube, diameters and flow rates as in (a), the velocity pressure factors for sections 1, 2 and 3 are 6.0, 3.0 and 10.0 respectively.

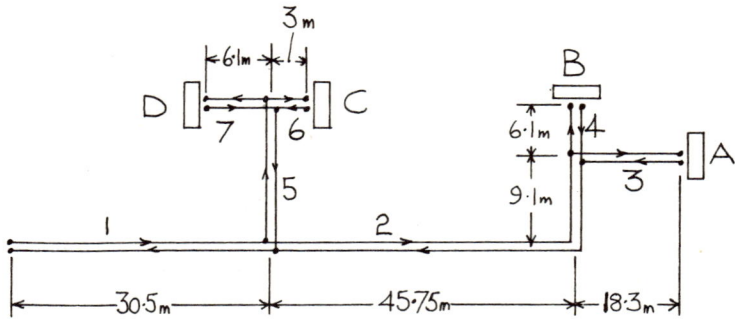

Fig. 4.8. Two-pipe hot-water heating system.

(Use tables for water at 75°C.)

Unit	A	B	C	D
Heat Emission, kW . . .	2.64	2.92	5.86	4.4

Ans.: (*a*) Allowing 25 per cent for pipe emission:

Section	1	2	3	4	5	6	7
Diameter, mm	32	20	15	15	20	20	15

(*b*) 31 kN/m² approx.

2. Some details of the index circuit of a two-pipe hot-water heating system are given in the table below. The mean temperature of the water is 75°C and the Medium Grade tubing is used throughout. Using the *I.H.V.E. Data Book*, determine the total resistance of the circuit.

Pipe section No.	Length flow and return, m	Flow rate, $\dot{m}$, kg/s	Pipe diameter, mm	Velocity pressure factor, K
1	122	12.6	100	8.0
2	61	5.92	80	5.6
3	91	1.37	50	7.8
4	152	0.832	40	8.4
5	122	0.252	25	10.3

(IHVE)

Ans.: 80 kN/m² approx.

5: High-Pressure Hot-Water Systems

Introduction

While the application of high-pressure hot water to space and process heating problems is probably the most outstanding development in heating engineering in recent years, its use for heating purposes is not a recent innovation. As early as 1831 a British patent was granted to Perkins, who developed a sealed-pipe system that operated with water temperatures up to 250°C. This early system consisted essentially of a natural circulation serving a number of pipe coils arranged in series in the heated rooms. The water was heated in a coil encased in a brickwork furnace. With the old Perkins system there was very little provision for controlling the temperature of the water, and as a result its use was limited to small buildings. It was superseded by the introduction of low-pressure hot water and the use of cast-iron sectional radiators for space heating.

The use of high-pressure hot water was re-introduced to this country from the Continent about 1930. It is particularly suitable for combined space and process heating; a typical temperature for these purposes being 180°C. To enable such high temperatures to be used the water must be kept above its saturation pressure. It will be seen from steam tables that if a water temperature of 180°C is required the pressure must be in excess of 1.0 MN/m²abs.

The simplest, but least practicable, method of pressurizing the system is by head tank, i.e. by the weight of a column of water. With this method it is important that only cool water below 100°C is allowed to expand out of the system into the head tank, which is open to the atmosphere via the vent and overflow pipes. A suitable arrangement is shown in Fig. 5.1, in which a vessel having a volume equal to the full expansion volume of the system is located at low level in the feed and expansion line. Since the feed and expansion pipe is connected into the main return pipe, only the cooler return water will expand out of the system and displace cold water from the low-level vessel into the high-level expansion vessel. The sparge pipe is included to encourage stratification and to prevent warm return water passing directly through the low-level expansion vessel. Another simple and inexpensive method is to use a sealed expansion vessel and allow the expansion volume of the water to create the required working pressure. The basic arrangement is shown in Fig. 5.2.

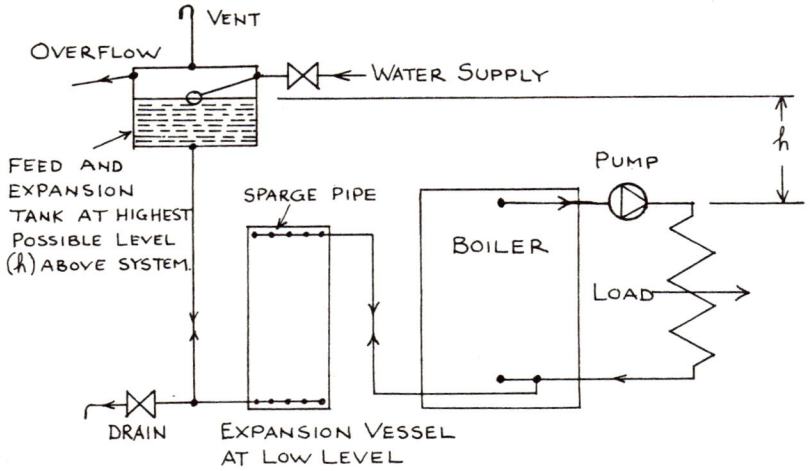

Fig. 5.1

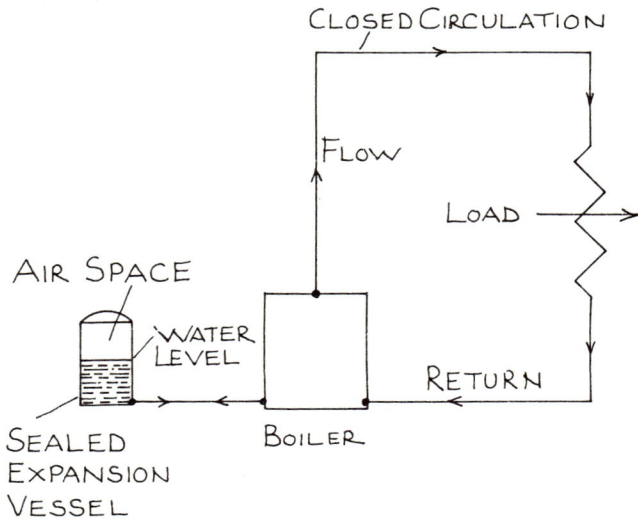

Fig. 5.2. Closed circulation, basic arrangement.

Consider the sealed expansion vessel shown in Fig. 5.3.

During initial filling an air volume (V_1) is sealed in the vessel when the water reaches the level shown in (a). Its pressure (p_1) at this stage will be atmospheric. As filling continues the water level rises and compresses the air. When the system is full, as shown in (b), the air volume has been

reduced from V_1 to V_2 and its pressure increased from p_1 to p_2, corresponding to the static head of the feed tank. On heating-up, the expansion volume
(E) of the water in the system enters the vessel and reduces the air volume
from V_2 to V_3 and increases its pressure from p_2 to p_3. Assuming that the
temperature of the air remains constant, then the compression of the air
will follow the law:

$$pV = C$$

and
$$p_1V_1 = p_2V_2 \tag{5.1}$$

also
$$p_2V_2 = p_3V_3 \tag{5.2}$$

since
$$V_3 = V_2 - E$$

then from Eq (5.2)
$$p_2V_2 = p_3(V_2 - E)$$
$$p_2V_2 = p_3V_2 - p_3E$$

Rearranging
$$p_3V_2 - p_2V_2 = p_3 . E$$

that is
$$V_2(p_3 - p_2) = p_3 . E$$

from which
$$V_2 = \frac{p_3 . E}{p_3 - p_2}$$

Substituting for V_2 in Eq (5.1)

$$p_1V_1 = \frac{p_2 . p_3E}{p_3 - p_2}$$

from which
$$V_1 = \frac{p_2p_3E}{p_1(p_3 - p_2)} \tag{5.3}$$

Alternatively, if compression of the gas is adiabatic:

$$pV^n = C$$

AIR PRESSURE $\left(P\right)$ AND VOLUME $\left(V\right)$:-

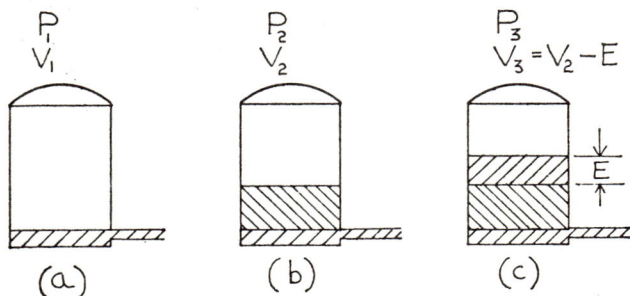

$\begin{array}{ccc} P_1 & P_2 & P_3 \\ V_1 & V_2 & V_3 = V_2 - E \end{array}$

(a) (b) (c)

Fig. 5.3. Sealed expansion vessel

and $\qquad\qquad p_1 V_1{}^n = p_2 V_2{}^n$ $\qquad$ (5.4)

also $\qquad\qquad p_2 V_2{}^n = p_3 V_3{}^n$ $\qquad$ (5.5)

since $\qquad\qquad V_3 = V_2 - E$

then from Eq (5.5) $\qquad p_2 V_2{}^n = p_3(V_2 - E)^n$

then by taking the nth root of both sides and simplifying

$$V_2 = \frac{p_3{}^{\frac{1}{n}} E}{p_3{}^{\frac{1}{n}} - p_2{}^{\frac{1}{n}}}$$

Substituting for V_2 in Eq (5.4),

$$p_1 V_1{}^n = \frac{p_2 p_3{}^{\frac{1}{n}} E}{p_3{}^{\frac{1}{n}} - p_2{}^{\frac{1}{n}}}$$

from which $\qquad\qquad V_1 = \dfrac{p_2{}^{\frac{1}{n}} p_3{}^{\frac{1}{n}} E}{p_1{}^{\frac{1}{n}}(p_3{}^{\frac{1}{n}} - p_2{}^{\frac{1}{n}})}$ $\qquad$ (5.6)

In equations (5.3) and (5.6)

p_1 = initial air pressure, normally atmospheric.

p_2 = air pressure with system full and cold. Due to static head of feed tank or pressure in filling main if used direct.

p_3 = final air pressure in vessel. Selected to support with safety the maximum desired working temperature at most vulnerable point in circulation.

V_1, V_2, V_3 = air volume at p_1, p_2, p_3 respectively.

E = total expansion volume of the water in the system.

n = 1.405 for air and 1.4 for nitrogen.

If $\qquad v_1$ = specific volume of water at t_1.

v_2 = specific volume of water at t_2.

and $\qquad Q$ = water content of system at t_1.

then $\qquad E = Q \cdot \dfrac{(v_2 - v_1)}{v_1}$

The change in water level in a cylindrical vessel may be found as follows:

Let $\qquad h$ = height of air space above level of water connection to system.

x = movement of water level for pressure change p_1 to p_2.

d = diameter of vessel.

Then for isothermal compression of the air:

$$\frac{V_1}{V_2} = \frac{p_2}{p_1}$$

since

$$V_1 = h\frac{\pi d^2}{4}$$

and

$$V_2 = h\frac{\pi d^2}{4} - x\frac{\pi d^2}{4}$$

$$\frac{h}{h-x} = \frac{p_2}{p_1}$$

from which

$$x = h\left[1 - \frac{p_1}{p_2}\right] \tag{5.7}$$

Similarly for adiabatic compression of the air:

$$x = h\left[1 - \left(\frac{p_1}{p_2}\right)^{\frac{1}{n}}\right] \tag{5.8}$$

Pressurization by Steam

Early types of modern systems use the steam in an ordinary steam boiler to provide the necessary pressure, the water at near saturation temperature being taken from the boiler instead of the steam. With this arrangement the natural relationship between boiling point and pressure determines the maximum flow water temperature permissible in the system. If the pressure is reduced due to changes in static head and frictional resistance, cavitation may occur. This can be avoided by mixing cool return water from the system with the hot water leaving the boiler. In this way the flow-water temperature can be adjusted to be at least 10°C, for safety, below the saturation temperature corresponding to the pressure prevailing at the most vulnerable point in the system. The basic arrangement is shown in Fig. 5.4.

As an alternative to using the boiler steam space, a separate small steam boiler may be used to supply steam to a sealed expansion vessel. A further alternative is to use a cascade type of water heater.

This type of heater is particularly suitable for heating large quantities of water to high temperature, and is generally used in association with steam plants designed primarily for power production. The basic arrangement is shown in Fig. 5.5. Return water from the heating circulation is passed into the top of the cascade heater, where it is pressurized and heated to near saturation temperature by direct contact with condensing steam as it cascades down a number of perforated trays. The condensate mixes with the water from the system and collects in the large water space at the bottom of the heater, from where it is taken to supply the heating circulation and the boiler.

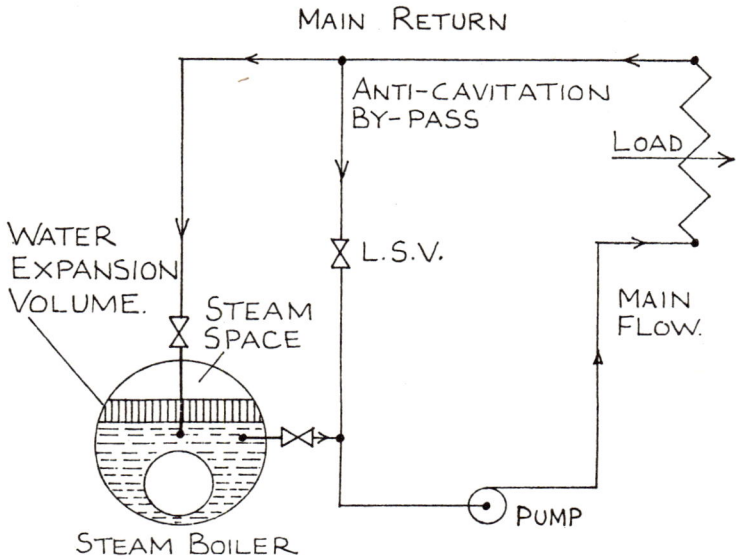

Fig. 5.4. Pressurization by steam, basic arrangement.

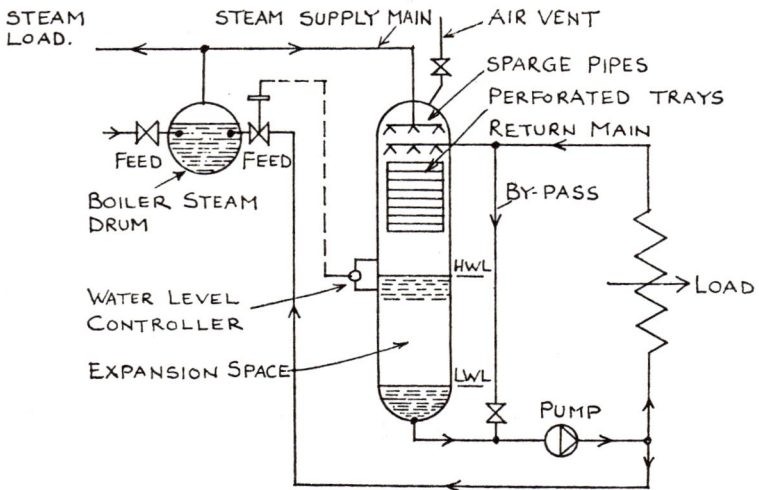

Fig. 5.5. Cascade Heater, Basic arrangement.

The heating system is thus supplied with condensate containing no dissolved solids and having definite corrosive tendencies. It is important therefore that the boiler feed water and any make-up water to the heater is correctly treated and deaerated to reduce the risk of corrosion. Cascade water heaters are ideal in cases where existing steam boilers are to be used and where steam and high-temperature waters are to be supplied from a common boiler plant. They are also used in high-temperature-water regenerative process-heating-power cycles.

Consider the simple arrangement shown in Fig. 5.6 when the pump is: (*a*) in the main flow at A, and (*b*) in the main return at B. Allow an anti-flash margin of 10°C and take the density of water at 900 kg/m³.

The steam pressure is assumed constant, and for this reason is adopted as a datum to which the pressure at other points in the system may be related. Pressure changes are considered to be either positive or negative if they are above or below datum respectively.

The following pressure distribution curves are shown in Fig. 5.7:

Curve 1. Pressure changes due to changes in position only.
Curve 2. Pressure changes due to frictional resistance with pump running in flow main at A.
Curve 3. Total pressure changes with pump running in flow main at A. That is, the algebraic sum of curves 1 and 2.
Curve 4. Pressure changes due to frictional resistance with pump running in return main at B.
Curve 5. Total pressure changes with pump running in return main at B. That is, the algebraic sum of curves 1 and 4.

With the pump running in the flow at A the point of minimum pressure is at the pump suction, that is:

$$800 - 41.5 = 758.5 \text{ kN/m}^2 \text{ abs.}$$

The boiling point for this pressure is found from steam tables to be approximately 168°C. Allowing a minimum anti-flash margin of 10°C, the maximum permissible flow-water temperature will therefore be 168−10=158°C.

Should the pump fail, the pressure throughout the system will revert to that fixed by curve No. 1. In these circumstances the minimum pressure is observed to be at *c* and *d*, that is:

$$800 - 114.7 = 685.3 \text{ kN/m}^2 \text{ abs.}$$

for which pressure the boiling point is 164°C. Since this is higher than the allowed maximum flow-water temperature by 164 − 158 = 6°C, cavitation will not occur.

With the pump running in the return at B the point of minimum pressure is at point *d*, that is:

$$800 - 199.7 = 600.3 \text{ kN/m}^2 \text{ abs.}$$

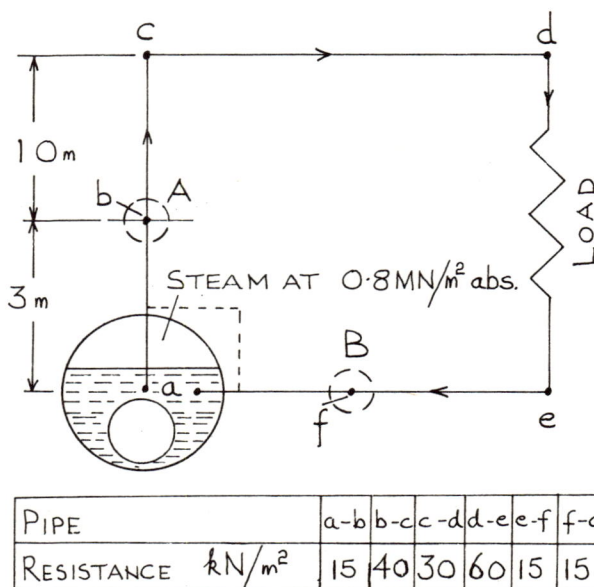

PIPE	a-b	b-c	c-d	d-e	e-f	f-a	TOTAL
RESISTANCE kN/m²	15	40	30	60	15	15	175

Fig. 5.6. Pressurization by steam, position of pump.

At which pressure the boiling point is approximately 159°C. Allowing a margin of 10°C, the maximum permissible flow-water temperature will therefore be 149°C. The flow-water temperature is found to be:

With pump in flow at A = 158°C
With pump in return at B = 149°C.

This shows that it is possible to operate with a higher flow-water temperature when the pump is located in the flow than with the pump in the return. In practice, the maximum temperature is sometimes fixed relative to the static pressure in the high-level mains, and would in this case be 164 − 10 = 154°C.

If the heat load on the system is, say, 6 MW and the temperature drop across the system is 50°C the amount of cool return water that should by-pass the boiler and mix with the hot water leaving the boiler will be found, for pump in flow at A, as follows:

Let x = by-pass at 158 − 50 = 108°C

then $1 − x$ = boiler water at 170°C

Flow temperature = Mixture temperature = 158°C

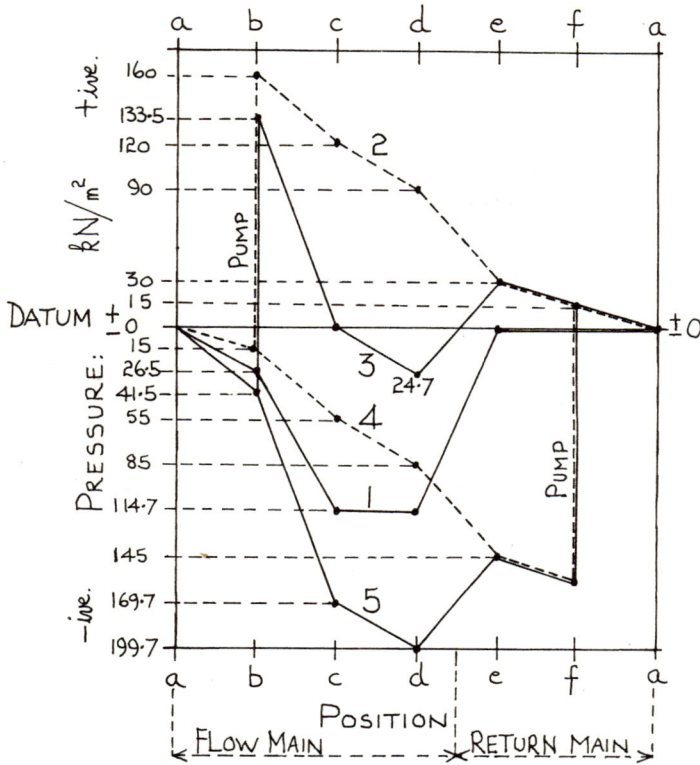

Fig. 5.7. Pressure distribution curves.

Using the specific enthalpy of the saturated liquid, kJ/kg (from steam tables), at these temperatures then by method of mixtures:

$$x \cdot 453 + (1-x)721 = 666$$

from which $\qquad x = 0.205$

that is $\qquad$ By-pass = 20.5 per cent

System flow rate

$$= \frac{6 \times 10^3}{666 - 453} = 28.2 \text{ kg/s}$$

Therefore by-pass

$$= 0.205 \times 28.2 = 5.78 \text{ kg/s}$$

and water flow rate leaving the boiler will be

$$28.2 - 5.78 = 22.42 \text{ kg/s}$$

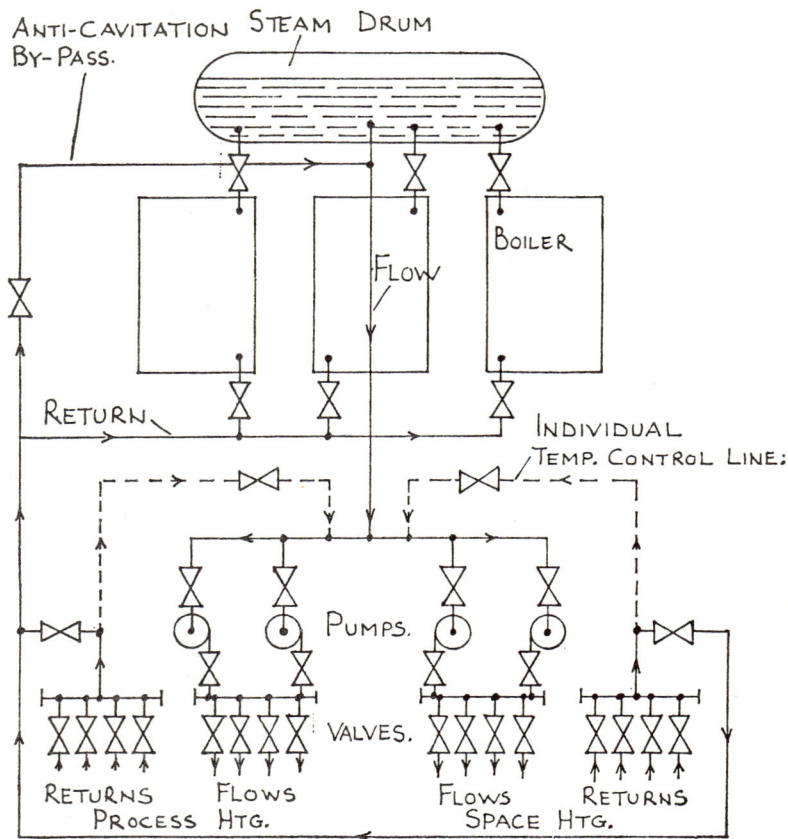

Fig. 5.8. Combined space and process heating system.

A combined space and process heating system is shown diagrammatically in Fig. 5.8. Each service has its own temperature-control link, and in addition there is a permanent by-pass that ensures that the temperature of the water leaving the boiler is always about 10°C lower than the boiling point corresponding to the pressure at the most vulnerable point in the system. With steam pressurization the maximum flow-water temperature is always below the saturation temperature of the water within the boiler.

Pressurization by Air or Nitrogen

There are several methods of gas pressurization in use, depending on whether all or part of the water expansion is to be accommodated within the sealed expansion vessel or whether it is required to operate with a constant or variable gas pressure. Assume that the sealed expansion vessel of a nitrogen pressurized hot-water heating system contains water initially at 120°C and has a 2.27 m³ expansion space above the water level and that the total pressure under these conditions is 1.82 MN/m² abs. i.e.

$$p = p_{N_2} + p_S = 1.82$$

where

p_{N_2} = partial pressure of the nitrogen
p_S = saturated vapour pressure.

From steam tables, the saturated vapour pressure corresponding to a water temperature of 120°C, is found to be 0.2 MN/m² abs. The nitrogen pressure required will therefore be:

$$1.82 - 0.2 = 1.62 \text{ MN/m}^2 \text{ abs.}$$

The amount of nitrogen in the expansion space may be determined as follows:

From the general gas equation ($pv = RT$)

$$v = \frac{RT}{p}$$

where

v = specific volume
T = absolute temperature
p = pressure
R = universal gas constant
  = 297 J/kg K

then

$$v = \frac{297(120 + 273)}{1.62 \times 10^6} = 0.072 \text{ m}^3/\text{kg}$$

The expansion space therefore contains:

$$\frac{2.27}{0.072} = 31.4 \text{ kg of N}_2$$

Assume that the water temperature is increased to 150°C and that the expansion space is reduced to 1.7 m³ due to expansion of water from the heating system. The specific volume of the nitrogen will now be:

$$v = \frac{1.7}{31.4} = 0.054 \text{ m}^3/\text{kg}$$

and its pressure will be:

$$p_{N_2} = \frac{297(150 + 273)}{0.054 \times 10^6} = 2.3 \text{ MN/m}^2$$

From steam tables, the saturated vapour pressure of water at 150°C is found to be 0.475 MN/m² abs. The total pressure in the expansion space will therefore be 2.3 + 0.475 = 2.775 MN/m² abs. If, alternatively, the initial pressure is to remain constant, then some nitrogen or some water must be released from the system as the water is heated from 120°C to 150°C. The amount of water that should be released is clearly 2.27 − 1.7 = 0.57 m³. Assume that no water is released, then the amount of nitrogen which should be released from the system may be found as follows:

Saturated vapour pressure at 150°C = 0.475 MN/m² abs.

The nitrogen pressure required to maintain a total pressure of 1.82 MN/m² abs will therefore be:

$$1.82 - 0.475 = 1.345 \text{ MN/m}^2 \text{ abs}$$

and the specific volume will be:

$$v = \frac{297(150 + 273)}{1.345 \times 10^6} = 0.0934 \text{ m}^3/\text{kg}$$

The amount of nitrogen is therefore

$$\frac{1.7}{0.0934} = 18.2 \text{ kg}$$

This means that 31.4 − 18.2 = 13.2 kg of nitrogen must be released from the system. If the water in the expansion vessel is only at ambient temperature its saturated vapour pressure is relatively small and may be neglected. Three basic arrangements only are described here.

In Fig. 5.9 is shown a common arrangement for accommodating only a small part of the water expansion within the system. The expansion volume from cold to minimum working temperature is allowed to escape to the spill tank, while the water expansion over the minimum to maximum temperature range is retained within the sealed expansion vessel. On cooling down, the gas pressure is reduced due to water shrinkage, and at some predetermined low pressure the feed pumps automatically start and return the original expansion volume back to the system. Since the spill-valve and feed-pump controls require considerable pressure differential to prevent simultaneous operation, the design pressure is higher than for a steam-pressurized system for the same flow-water temperature. A typical comparison would be as follows:

Steam system:

Operating pressure required to support maximum
   flow water temperature   1.05 MN/m²
Safety valve setting   1.1 MN/m²

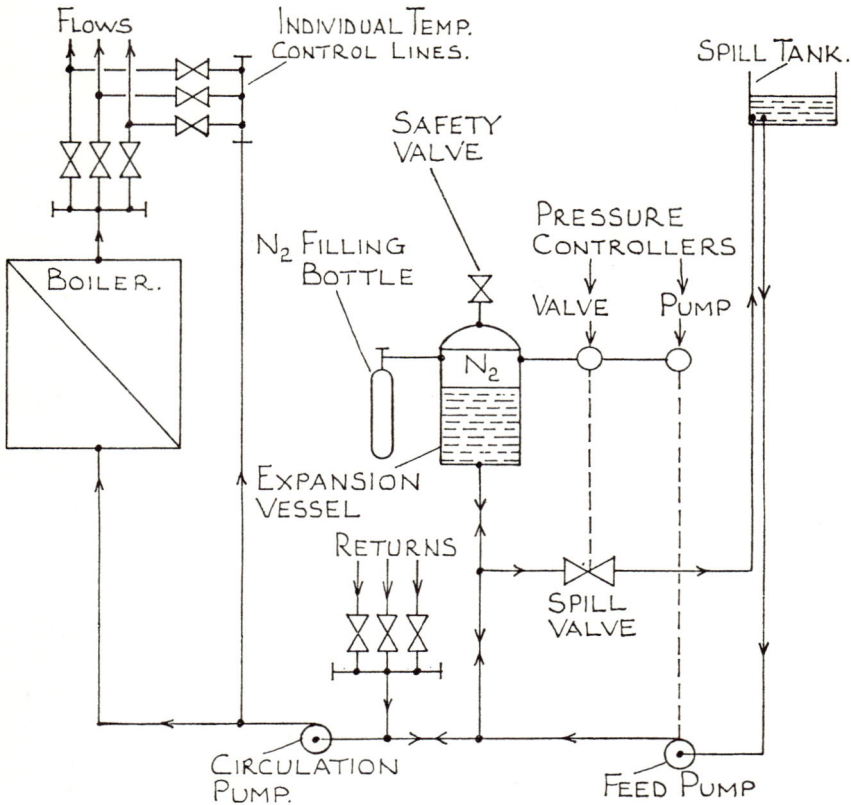

Fig. 5.9. Pressurization by gas with water expansion passed to spill tank.

Gas system:

Operating pressure as for steam system	1.05 MN/m^2
Pressure differential of spill valve and feed pump controller	0.15 MN/m^2
Safety valve setting	1.25 MN/m^2

The system is normally pressurized when cold from the nitrogen filling bottles or, alternatively, by compressed air.

An arrangement for retaining the full water expansion within the system and operating at constant pressure is shown in Fig. 5.10. The system is pressurized to the full design pressure when cold. The pressure increases as the system is heated up, and at some pre-determined high pressure the relief valve opens and allows nitrogen to escape to the low-pressure receiver, from where it is taken to the compressor, recompressed and passed to the

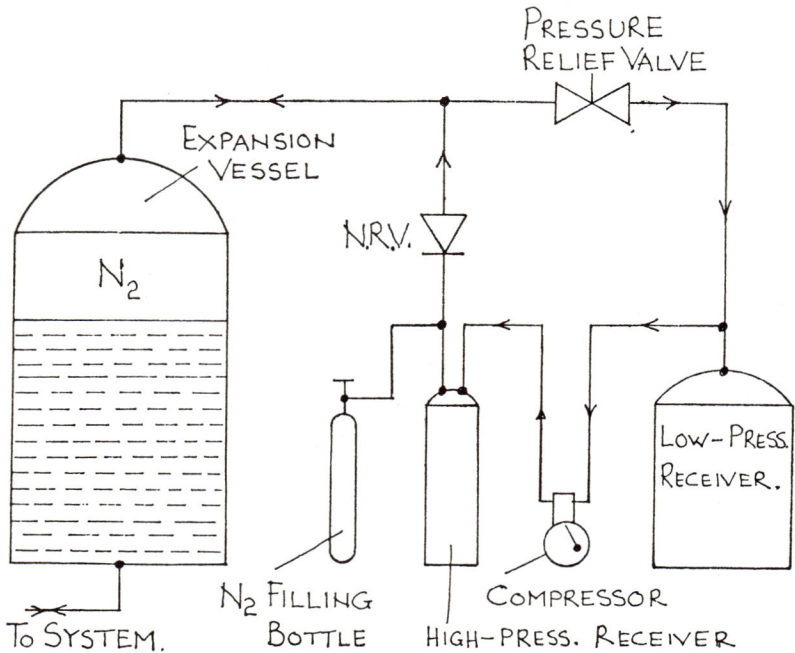

Fig. 5.10. Full expansion retained within system; constant pressure N_2.

high-pressure receiver ready for maintaining pressure on the system as the water temperature falls.

An alternative to the arrangement shown in Fig. 5.10 uses compressed air as shown in Fig. 5.11. The system operates at constant pressure and retains the full water expansion within the sealed expansion vessel. With this system air is released as the water level in the vessel rises and replenished from the compressor as the water level falls on contracting.

With systems that use air the corrosion hazards due to oxygenation of the water may be reduced by: (*a*) using a float on the water surface, thus reducing the contact area; (*b*) heating the water to near boiling point; or (*c*) a layer of paraffin as a barrier between the water and the air.

Example 5.1. A hot-water heating system is filled with 2 m^3 of water at 10°C from a tank located 15 m above the air vessel. The system is connected to a sealed expansion vessel and is required to operate with flow and return water temperatures of 120°C and 80°C respectively. Assuming a level site and neglecting resistance changes, size the air vessel. Assume constant air temperature.

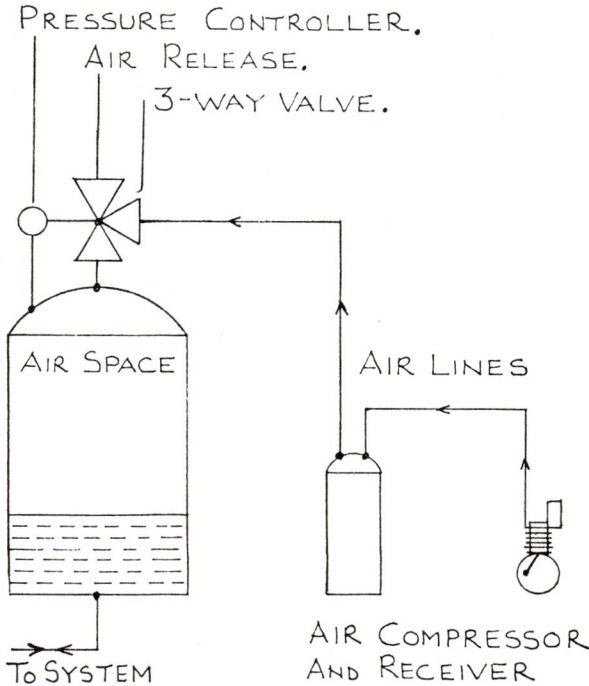

Fig. 5.11. Full expansion with constant air pressure.

Specific volume of water at 100°C = 0.00104 m³/kg
Specific volume of water at 10°C = 0.00100 m³/kg
Anti-flash margin   = 15°C

For constant air temperature, use Eq (5.3).

$$V_1 = \frac{p_2 p_3 E}{p_1 (p_3 - p_2)}$$

$$E = 2\left(\frac{104 - 100}{100}\right) = 0.08 \text{ m}^3$$

p_1 = atmospheric = 0.1 MN/m² abs.

$$p_2 = \left(\frac{15 \times 9.8}{0.001 \times 10^6}\right) + 0.1 = 0.247 \text{ MN/m}^2 \text{ abs.}$$

p_3 should be selected from steam tables for a temperature of
120 + 15 = 135°C

i.e. p_3 = 0.314 MN/m² abs.

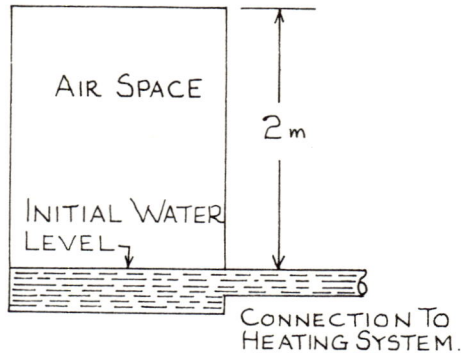

Fig. 5.12. Expansion vessel.

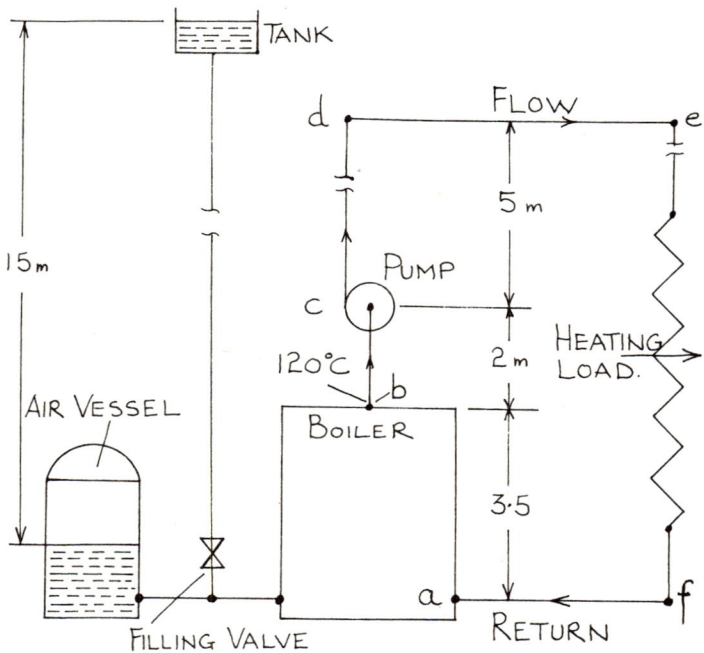

Fig. 5.13. Air pressurization with pump in flow main.

Then $\quad V_1 = \dfrac{0.247 \times 0.314 \times 0.08}{0.1(0.314 - 0.247)} = 0.93 \text{ m}^3$

This is the volume of air that should be initially sealed in the vessel above the level of the pipe connection to the system. Assuming a height of 1.8 m, the diameter of the vessel would be:

$$\frac{1.8\,\pi d^2}{4} = 0.93$$

from which

$$d = 0.8 \text{ m approx.}$$

Example 5.2. The expansion vessel of a hot-water heating system is shown in Fig. 5.12. The normal operating range of the system is between 0.2 and 0.4 MN/m^2 abs. Calculate the changes in water level between these pressures. Assume isothermal conditions.

The rise in water level as the pressure increases from atmospheric to 0.2 MN/m^2 abs will be, from Eq (5.7):

$$x = 2\left[1 - \frac{0.1}{0.2}\right] = 1 \text{ m}$$

Similarly, to 0.4 MN/m^2 abs

$$x = 2\left[1 - \frac{0.1}{0.4}\right] = 1.5 \text{ m}$$

The change in water level between 0.2 and 0.4 MN/m^2 abs is therefore:

$$1.5 - 1 = 0.5 \text{ m}$$

Example 5.3. The hot-water system shown in Fig. 5.13 operates at a mean temperature of 110°C. The system is filled with 4.5 m^3 of water at 10°C from a tank located 15 m above the air vessel.

(*a*) Determine the size of the air vessel. Assume constant air temperature.

Pipe Section	*a–b*	*b–c*	*c–d*	*d–e*	*e–f*	*f–a*	Total
Resistance, (Δp) kN/m^2	3	3	6	12	100	6	130

(*b*) If the diameter of the air vessel is 1.5 m, calculate the changes in water level that occur as the water is heated from cold to operating temperature.

Data:

Specific volume of water at 10°C $= 0.00100 \text{ m}^3/\text{kg}$
Specific volume of water at 110°C $= 0.00105 \text{ m}^3/\text{kg}$
Flow water temperature = 120°C. Anti-flash margin, 10°C

(*a*) From the total pressure distribution curve, Fig. 5.14, it will be seen that the point of minimum pressure, when the pump is running, is at the pump suction.

The pressure at this point should be sufficient to support a temperature of $120 + 10 = 130°C$, and from steam tables is found to be 0.27 MN/m^2 abs. The datum pressure is 50.1 kN/m^2 above this, and is equal to:

$$0.27 + 0.0501 = 0.32 \text{ MN/m}^2$$

If the pump fails the lowest pressure will be at level $d-e$ and will be

$$0.32 - \frac{9.8(3 + 2 + 5)}{0.00105 \times 10^6} = 0.23 \text{ MN/m}^2 \text{ abs.}$$

Boiling point for this pressure = $125°C$.

Since the maximum flow water temperature is $120°C$, there will be $125 - 120 = 5°C$ margin should the pump fail. This margin is ample for safety, therefore size the air vessel to give a pressure of 0.32 MN/m^2 when the system is hot.

$$p_2 = \left(\frac{15 \times 9.8}{0.001 \times 10^6} \right) + 0.1 = 0.25 \text{ MN/m}^2$$

$$E = 4.5 \left(\frac{105 - 100}{100} \right) = 0.225 \text{ m}^3$$

$$V_1 = \frac{0.25 \times 0.32 \times 0.225}{0.1(0.32 - 0.25)} = 2.57 \text{ m}^3$$

$$h = \frac{2.57 \times 4}{1.5^2 \pi} = 1.45 \text{ m}$$

(*b*) The rise in water level as the pressure increases from atmospheric to 0.25 MN/m^2 abs will be, from Eq (5.7),

$$x = 1.45 \left[1 - \frac{0.1}{0.25} \right] = 0.87 \text{ m}$$

Similarly, to 0.32 MN/m^2 abs

$$x = 1.45 \left[1 - \frac{0.1}{0.32} \right] = 1 \text{ m}$$

The change in water level between 0.25 and 0.32 MN/m^2 abs will therefore be:

$$1 - 0.87 = 0.13 \text{ m}$$

Example 5.4. A high-pressure hot-water system arranged for steam pressurization operates at a boiler pressure of 0.9 MN/m^2 abs and serves heaters installed 12 m above the water level of the boiler. The pump is installed in the flow main 3 m above the boiler water level, and develops 0.18 MN/m^2. The frictional resistance between the boiler and the pump, and between the pump and the heaters, is 0.02 and 0.05 MN/m^2 respectively.

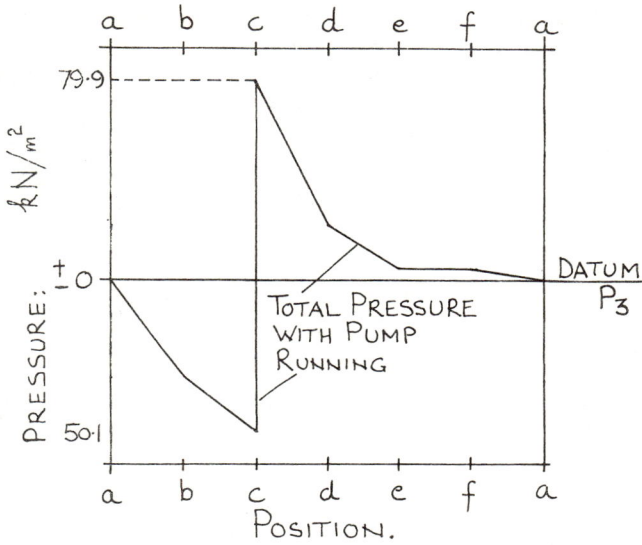

Fig. 5.14. Pressure distribution with pump running.

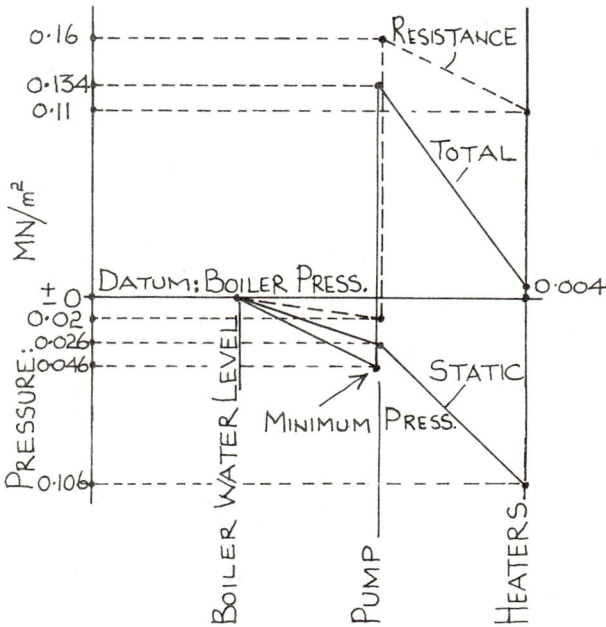

Fig. 5.15. Pressure distribution, steam pressurization.

(*a*) Give a diagram to show the position, resistance and total pressure changes between the boiler and the heaters.

(*b*) Determine the maximum permissible flow-water temperature when the pump is running. Assume an anti-flash margin of 10°C and a water density of 900 kg/m³.

(*c*) If the total heat supplied is 4.4 MW and the temperature drop across the system is 55°C, calculate the amount of return water that should by-pass the boiler.

(*a*) See Fig. 5.15.

(*b*) The point of minimum pressure when the pump is running is, from the total pressure curve of Fig. 5.15, at the pump inlet and will be:

$$0.9 - 0.046 = 0.854 \text{ MN/m}^2 \text{ abs.}$$

The evaporation temperature at this pressure is found from steam tables to be 173°C. The maximum permissible temperature will therefore be:

$$173 - 10 = 163°C$$

(*c*) Let x = by-pass at $163 - 55$ = 108°C
then $1 - x$ = boiler water at (from steam tables)   = 175°C
        Flow temperature = mixture temperature = 163°C

Using the specific enthalpy of the saturated liquid, kJ/kg (from steam tables), at these temperatures then by method of mixtures:

$$453x + 743(1 - x) = 682$$

from which $x = 0.21$, i.e. 21 per cent

System flow rate

$$= \frac{4\,400}{682 - 453} = 19.2 \text{ kg/s}$$

Therefore, by-pass

$$= 0.21 \times 19.2 = 4.03 \text{ kg/s}$$

Example 5.5. The steam pressurized system shown in Fig. 5.16 is required to operate with a flow-water temperature of 180°C at the heaters and an anti-flash margin of 10°C when the circulating pump is off. Determine:

(*a*) the minimum steam pressure;
(*b*) the anti-flash margin at the heaters when the pump is running.

Pipe section	a–b	b–c	c–d	d–e	e–f	f–a	Total
Resistance, kN/m²	5	10	3	10	2	5	115

Take the density of water at 900 kg/m³.

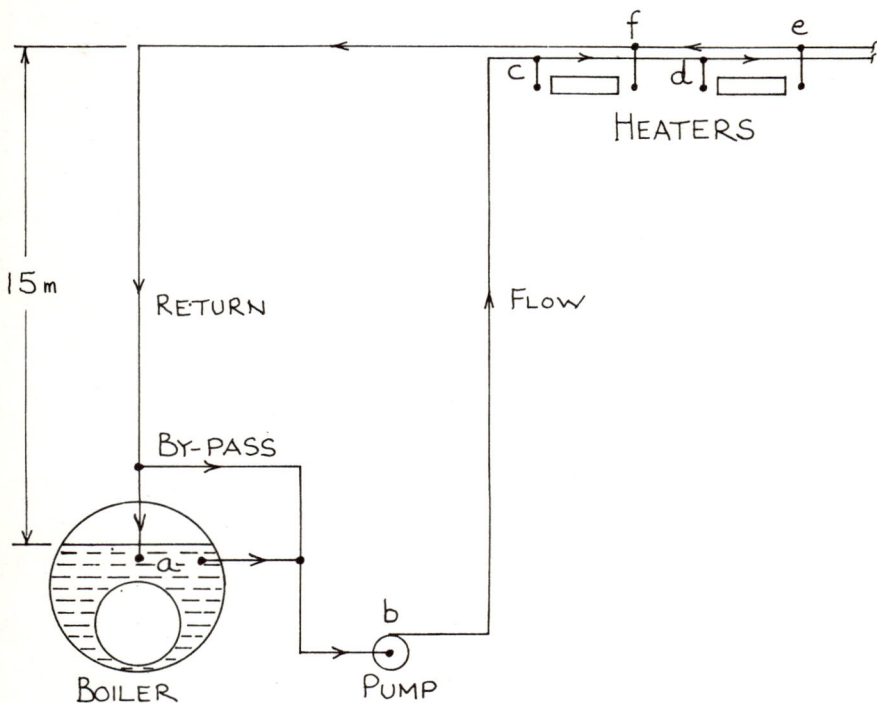

Fig. 5.16. Unit heaters at high level.

(*a*) When the pump is off the lowest pressure will be at level $c-d$, that is $15 \times 900 \times 9.8 \times 10^{-6} = 0.13 \, \text{MN/m}^2$ lower than the pressure of the steam in the boiler. At this level a pressure corresponding to an evaporation temperature of $180°\text{C} + 10°\text{C} = 190°\text{C}$ is required, and from steam tables is found to be $1.25 \, \text{MN/m}^2$ abs. The minimum steam pressure will therefore be:

$$1.25 + 0.13 = 1.38 \, \text{MN/m}^2 \text{ abs.}$$

(*b*) When the pump is running the pressure at point c will be:

Steam pressure + Pump pressure − Elevation change − Resistance $a-c$

$$= 1.38 + 0.115 - 0.13 - (0.005 + 0.1) = 1.35 \, \text{MN/m}^2 \text{ abs.}$$

Evaporation temperature at this pressure = $193°\text{C}$
Flow water temperature at c = $180°\text{C}$
Therefore anti-flash margin at c = $\underline{13°\text{C}}$

Example 5.6. A steam pressurized hot-water unit heater system is shown diagrammatically in Fig. 5.17. If individual room thermostats are used to provide simple on/off control of each unit heater fan motor determine the maximum flow water temperature that can be used when the pump is running: (i) at A; (ii) at B.

Data:

Number of unit heaters	60
Heat output per unit with fan on	35 kW
Heat output per unit with fan off	3 kW
Temperature difference at full load between flow and return water	55°C
Steam pressure in boiler	0.9 MN/m² abs
Water density to be taken at	900 kg/m³
Anti-flash margin	10°C
Specific heat capacity of water	4.2 kJ/kg°C

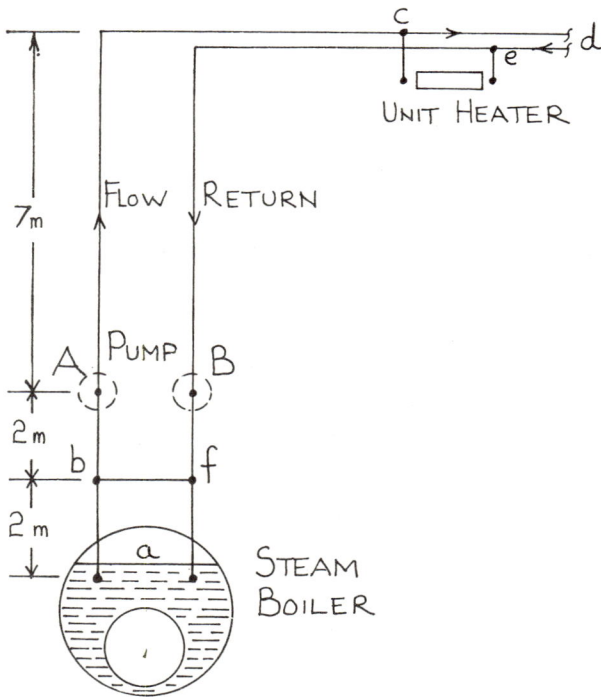

Fig. 5.17. Steam pressurization, alternative pump positions.

When the fan motors are switched off the heat output of the units is reduced and, since the system flow rate remains constant, the temperature drop across the units will also be reduced, with the result that the temperature of the water in the return main will be increased. This should be taken into account when determining the maximum flow water temperature for the system. Consider first the conditions in the return main at *e* when all fan motors are "off".

Pipe section	*a–b*	*b–A·*	*A–c*	*c–d–e*	*e–B*	*B–f*	*f–a*	Total
Resistance, kN/m^2 ..	5	5	5	85	5	5	5	115
Pipe emission, kW ..	NIL	NIL	23	175	12	NIL	NIL	210

The total pressure distribution is given in Fig. 5.18, from which the pressure at "e" in the return will be:

(i) With pump at A:

$$0.9 - 0.082 = 0.818 \text{ MN/m}^2$$

Evaporation temperature at this pressure	= 171°C
Anti-flash margin	= 10
Maximum permissible temperature at "*e*" in return	= 161°C

(ii) With pump at B:

$$0.9 - 0.197 = 0.703 \text{ MN/m}^2$$

Evaporation temperature at this pressure	= 165°C
Anti-flash margin	= 10
Maximum permissible temperature at "*e*" in return	= 155°C

Total connected load = 60 × 35 =	2 100
Pipes A–*c* =	23
c–d–e =	175
e–B =	12
	2 310 kW

$$\text{System flow rate} = \frac{2\,310}{55 \times 4.2} = 10 \text{ kg/s}$$

This should strictly be determined from knowledge of the change in specific enthalpy of the water for a 55°C change in temperature, but since the temperature level is not yet known, the temperature difference is used. The error involved is not likely to be more than 2 per cent.

Load with all fans "off":

Units, 60 × 3	= 180
Pipes A–*e*, 23 + 175	= 198
	378 kW

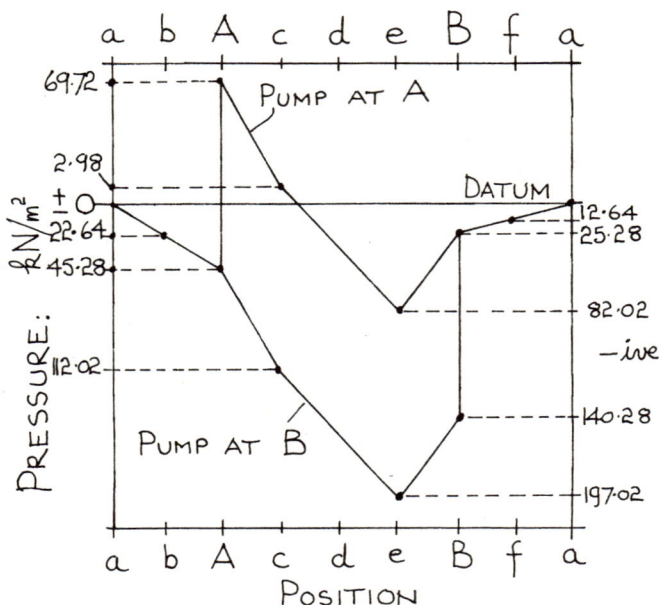

Fig. 5.18. Pressure distribution, alternative pump positions.

The maximum permissible flow water temperature will therefore be:

(i) With pump at A:

$$161 + \frac{378}{10 \times 4.2}$$

$$= 161 + 9 = 170°C$$

(ii) With pump at B:

$$= 155 + 9 = 164°C$$

These temperatures may, however, be excessive under normal operating conditions.

(i) With pump at A.

Pressure at pump suction = 0.9 − 0.045 = 0.855 MN/m²
Evaporation temperature at this pressure = 173°C
Anti-flash margin   = 10

Maximum permissible temperature   = 163°C

If the flow-water temperature had been fixed at 170°C the margin at the pump suction would be only 173 − 170 = 3°C, which is unsatisfactory.

The maximum permissible flow water temperature with the pump at A will therefore be 163°C and the return water temperature, when all fans are "off", will be $163 - 9 = 154°C$. The margin at e will be $171 - 154 = 17°C$. Should the pump fail, the minimum pressure in the flow main will be at c, i.e. $0.9 - (2 + 2 + 7)900 \times 9.8 \times 10^{-6} = 0.8$ MN/m^2, at which pressure the evaporation temperature is found from steam tables to be 170°C, which is well above the maximum flow temperature of 163°C.

(ii) With pump at B.

The pressure in the flow main at c will be $0.9 - 0.112 = 0.788$ MN/m^2.

Evaporation temperature at this pressure	= 170°C approx.
Anti-flash margin	= 10
Maximum permissible temperature	= 160°C

In this case the pressure at c will increase if the pump fails, so there is no danger of flash steam occurring in the flow main. The above analysis show that the maximum flow water temperature should be limited to:

(i) With pump in flow main at A   = 163°C
(ii) With pump in return main at B = 160°C

Example 5.7. The canteen of a factory centre is to be supplied with steam at 110°C from a heat exchanger using high-pressure hot-water as the primary heating medium.

(*a*) Recommend suitable primary flow and return water temperatures.

(*b*) If only the latent heat in the steam is used and only 70 per cent of the condensate is returned for re-use, calculate the load on the heat exchanger. Cooking load: 73 kW

(*c*) Determine the heating surface area. Take the thermal transmittance of the tubes at 680 W/m^2°C.

(*d*) Determine the flow rate through the primary circulation and recommend a pipe diameter.

(*e*) Give a sketch of a typical heat exchanger showing all essential features and control equipment.

(*a*) The minimum primary return temperature should be at least 10°C above the steam temperature, i.e. return temperature = $110 + 10 = 120°C$. Allowing a temperature drop of say 45°C across the heat-exchanger tubes, the flow-water temperature will be:

$$120 + 45 = 165°C$$

(*b*) From steam tables for steam at 110°C:

Pressure	= 0.14 MN/m^2
Specific enthalpy of saturated liquid	= 458 kJ/kg
Specific enthalpy of saturated vapour	= 2 690 " "
Latent heat	= 2 232 " "

Quantity of steam required

$$= \frac{73}{2\,232} = 0.033 \text{ kg/s}$$

Condensate returned for re-use = 0.7 × 0.033 = 0.023 kg/s

Assuming that this condensate would be returned to an open hot well, its temperature would be at, say, no more than 98°C.

Make-up water = 0.033 − 0.023 = 0.01 kg/s at, say, 10°C

Therefore feed water temperature will be

$$\frac{0.023 \times 98 + 0.01 \times 10}{0.033} = 71°C$$

Load on heat exchanger = 0.033 kg/s from feed water at 71°C to steam at 110°C, that is:

$$0.033\,[2\,232 + 4.2(110 - 71)] = 79 \text{ kW}$$

(c) Neglecting the rise in feed water temperature the temperature changes involved are:

Flow	165°C →	Return 120°C
Steam	110°C →	110°C
$\Delta t_1 =$	55°C	$\Delta t_2 =$ 10°C

$\Delta t_1 > 1.5 \Delta t_2$ therefore use the logarithmic temperature difference (LMTD)

$$\text{LMTD} = \frac{55 - 10}{\log_e \dfrac{55}{10}} = 26.4°C$$

Then:

$$\text{Heating surface area} = \frac{79 \times 10^3}{680 \times 26.4} = 4.3 \text{ m}^2$$

(d) Flow rate through primary circuit

$$= \frac{79}{4.2(165 - 120)} = 0.42 \text{ kg/s}$$

Allowing a water velocity of 1−2 m/s, the diameter from pipe-sizing tables will be 25 mm.

(e) A typical heat exchanger is shown in Fig. 5.19.

Example 5.8. The high-pressure hot-water heating system shown in Fig. 5.20 is required to operate at a constant nitrogen pressure of 0.8 MN/m² abs. Determine the maximum flow-water temperature that can

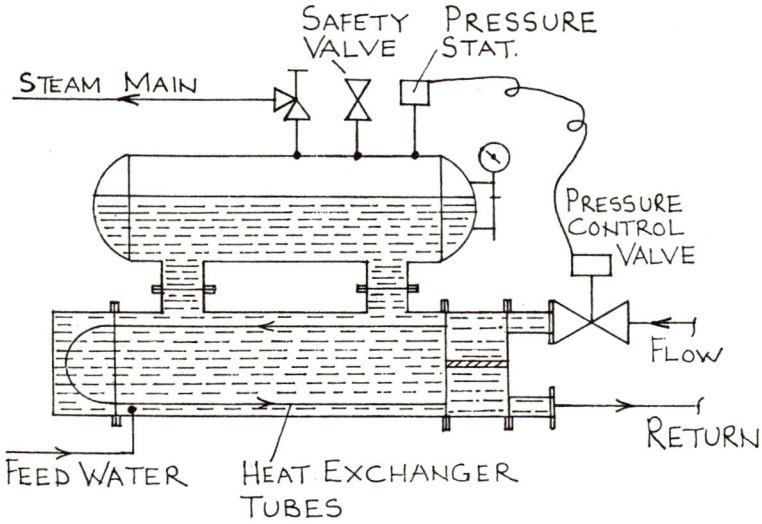

Fig. 5.19. Steam generator.

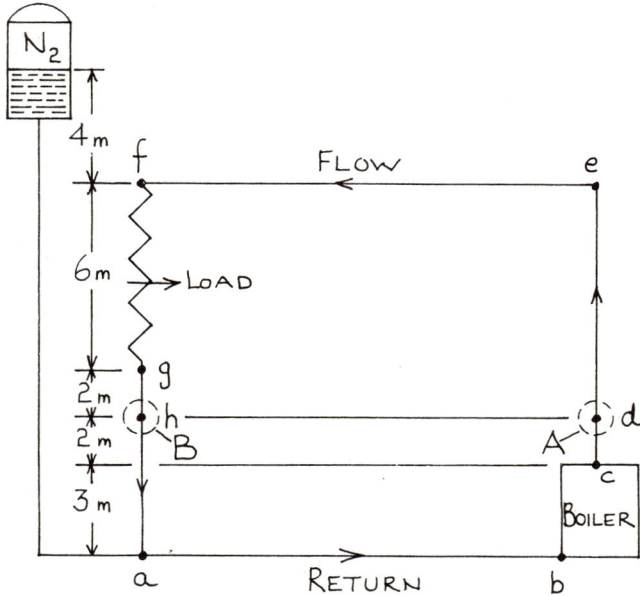

Fig. 5.20. High-level nitrogen pressure vessel.

be safely maintained if the pump is installed: (*a*) in the flow main at A, and (*b*) in the return main at B.

Section	a–b	b–c	c–d	d–e	e–f	f–g	g–h	h–a	Total
Resistance, kN/m²	25	18	18	25	35	180	12	12	325

Data:

Anti-flash margin   10°C

Density of water at mean temperature of system =   900 kg/m³
Density of water in expansion and feed pipe   = 1 000 kg/m³

Since the nitrogen pressure is constant the pressure at "*a*" will be

$$0.8 + (4 + 6 + 2 + 2 + 3)1\,000 \times 9.8 \times 10^{-6} = 0.97 \text{ MN/m}^2 \text{ abs.}$$

To find points of minimum pressure, draw diagram Fig. 5.21.

(*a*) *Pump in flow main at A:*

Minimum pressure is at the inlet to the pump and will be:

$$0.97 - 0.105 = 0.865 \text{ MN/m}^2 \text{ abs}$$

Evaporation temperature at this pressure = 174°C
Anti-flash margin   =   10
∴ Maximum temperature   = 164°C

(*b*) *Pump in return main at B:*

The minimum pressure in the flow main is at "*f*", and will be:

$$0.97 - 0.236 = 0.734 \text{ MN/m}^2 \text{ abs.}$$

Evaporation temperature at this pressure = 167°C
Anti-flash margin   =   10
∴ Maximum temperature   = 157°C

If the pump should fail the minimum pressure will be at *e–f*, that is

$$0.97 - 0.115 = 0.855 \text{ MN/m}^2 \text{ abs.}$$

Evaporation temperature at this pressure = 173°C

The margin in part (*a*) would then be reduced to 173 − 164°C = 9°C, which is still satisfactory, and in part (*b*) would be increased to 173 − 157 = 16°C.

Example 5.9. Using the data listed below, determine for the system shown in Fig. 5.22 the nitrogen pressure that should be applied if the pressure vessel is located:

(i) at A, and (ii) at B.

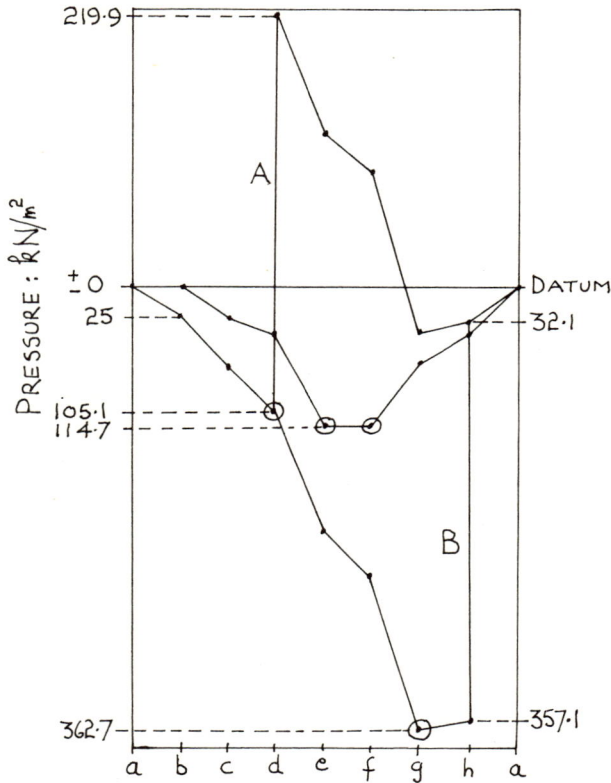

Fig. 5.21. Pressure distribution with vessel at high level.

Section	a–b	b–c	c–d	d–e	e–f	f–g	g–a	Total
Resistance, kN/m^2 ...	10	10	20	90	20	10	10	170

Data:

Flow-water temperature at inlet to the pump	= 180°C
Anti-flash margin	= 10°C
Density of water at mean temperature of system	= 900 kg/m^3
Density of water in feed and expansion pipe	= 1 000 kg/m^3

The pressure at the inlet to the pump should be sufficient to support an evaporation temperature of 180 + 10 = 190°C, and is found from steam tables to be 1.25 MN/m^2 abs. Using this as datum, the pressure distribution will be as shown in Fig. 5.23.

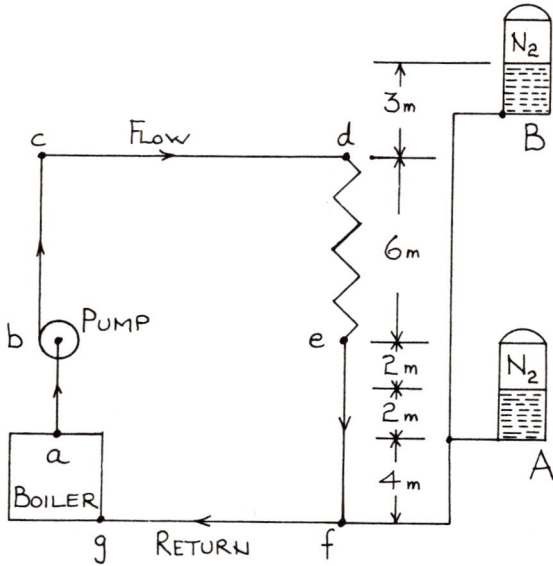

Fig. 5.22 Alternative pressure-vessel position.

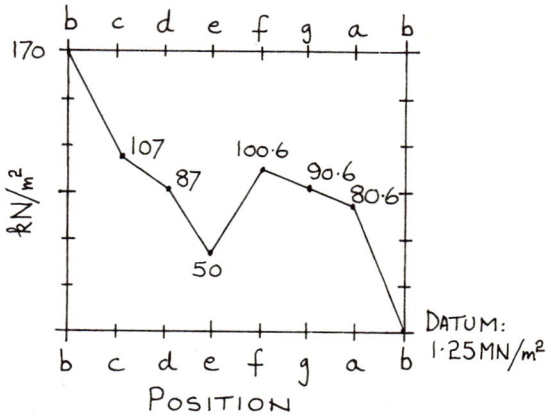

Fig. 5.23. Pressure distribution diagram.

The pressure at "*f*"

$$= 1.25 + 0.1 = 1.35 \text{ MN/m}^2$$

This is imposed by the nitrogen pressure plus the pressure due to the position of the pressure vessel above "*f*". That is, with the vessel at A:

$$1.35 = (6 \times 1\,000 \times 9.8 \times 10^{-6}) + N_2 \text{ pressure}$$

from which, N_2 pressure = 1.29 MN/m^2 abs.

and with the vessel at B:

$$1.35 = (16 \times 1\,000 \times 9.8 \times 10^{-6}) + N_2 \text{ pressure}$$

from which, N_2 pressure = 1.19 MN/m^2 abs.

Should the pump fail, the lowest pressure will be a level $c-d$ and equal to

$$1.35 - (14 \times 900 \times 9.8 \times 10^{-6}) = 1.23 \text{ MN/m}^2 \text{ abs.}$$

The evaporation temperature at this pressure = 189°C. Since the flow-water temperature is limited to 180°C, the margin under static conditions will be 9°C, which is satisfactory.

Problems

1. A medium-pressure hot-water heating system is connected to a sealed air vessel. Using the data given below and assuming isothermal conditions calculate:

(*a*) the volume of the air vessel;

(*b*) the change in the water level that occurs in the air vessel as the water is heated from filling to working temperature.

Data:

Flow water temperature	= 121°C
Height of feed tank above air vessel	= 15.24 m
Diameter of air vessel	= 1.22 m
Water expansion volume	= 0.23 m^3
Anti-flash margin	= 17°C

Ans.: (*a*) 2.04 m^3; (*b*) 1.22 m.

2. A hot-water system is pressurized by the arrangement shown in Fig. 5.24. Determine the effective area of the bellows.

Data:

Pressure within expansion bellows	= 0.93 MN/m^2
Mass acting on lever	= 454 kg.

Ans.: 0.014 m^2.

3. Using the data listed below and shown in Fig. 5.25, determine the steam pressure which must be maintained and the percentage of water which must be re-circulated via the by-pass in order to maintain about 11°C anti-flash margin at the heater when the pump is running. The lengths given include an allowance for local resistances.

Data:

Pressure drop in pipe work = 490.2 N/m² per metre length
       ”   ” ” ” heater =   5 980 N/m²
       ”   ” ” ” boiler = 24 920 N/m²
   Water density   = 930 kg/m³

Ans.: 0.42 MN/m² and 19% (approx.).

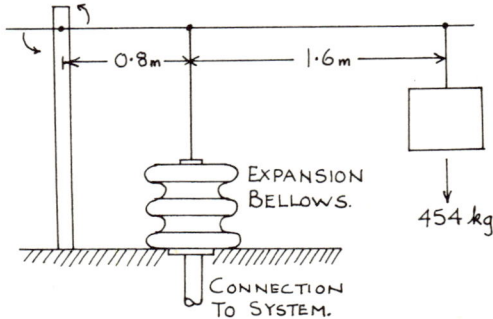

Fig. 5.24.

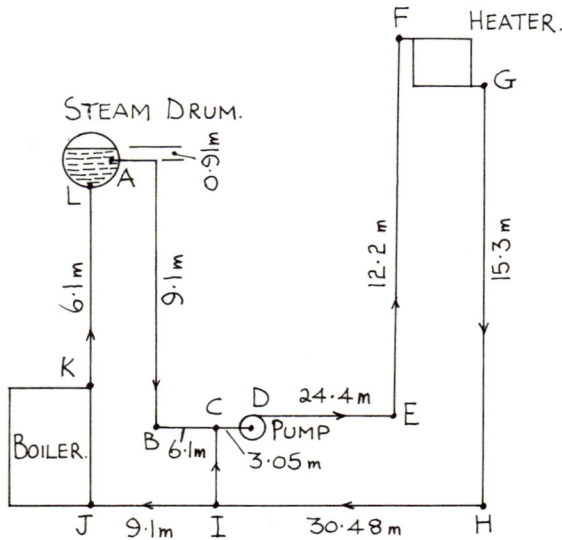

Fig. 5.25.

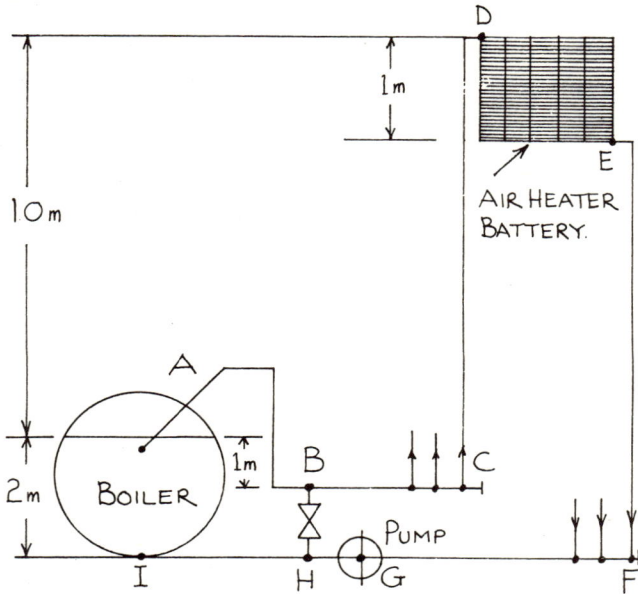

Fig. 5.26. Medium-pressure system.

4. A medium-pressure hot-water heating plant is arranged, as shown diagrammatically in Fig. 5.26, for operation at 0.4 MN/m^2 abs from the boiler steam space. If the pressure losses are 9 kN/m^2 through each section of pipe between the lettered junctions, 3 kN/m^2 through the boiler and 36 kN/m^2 through the air heater battery:

(*a*) Draw a diagram showing the total pressure changes around the system.

(*b*) Discuss the operational difficulties that might arise if a diverter valve were to be fitted, for thermostatic control purposes, between the flow and return connections of the heater battery.

Ans.: (*a*) Pressure distribution should be related to steam pressure.

(*b*) Anti-flash margin at E would be considerably reduced.

5. A high-pressure hot-water heating system is shown diagrammatically in Fig. 5.27. Using the data listed below, determine the maximum flow-water temperature that can be maintained if the pump is located: (*a*) in the flow pipe at A, and (*b*) in the return pipe at B. Allow a 11°C anti-flash margin in each case.

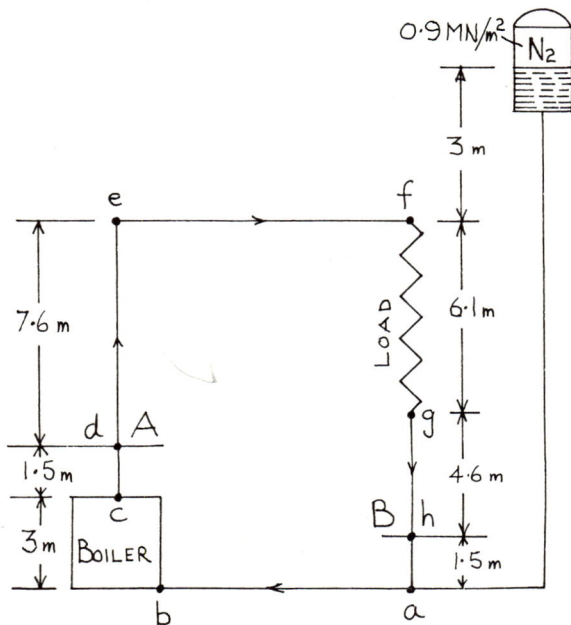

Fig. 5.27. Nitrogen vessel at high level.

Section	a–b	b–c	c–d	d–e	e–f	f–g	g–h	h–a	Total
Resistance, kN/m^2	14.1	5.8	8.7	14.4	14.4	57.8	14.4	14.4	144

Density of water = 900 kg/m^3

Nitrogen pressure constant at 0.9 MN/m^2 abs.

Ans.: (*a*) 170°C; (*b*) 165°C (nearest 5°C).

6: Pumps

Introduction

This chapter deals with centrifugal pumps of the type used for circulating water in closed heating and hot-water supply circuits. With these systems there is no static lift, and once the water is in motion no further pressure is required to keep it moving except for that required to overcome frictional resistance. The pump, therefore, must develop a pressure equal to that due to frictional resistance in the circuit. Manufacturers test pumps in open circuits with cold water and issue performance curves relating pump capacity in l/s* to pressure produced measured in N/m^2 for a given pump speed.

For a pump connected to a given piping circuit or working against a given equivalent orifice it has been shown that:

Quantity (Q) is proportional to speed (n)
Pressure (p) is proportional to the square of the speed (n)
Power (P) is proportional to the cube of the speed (n)

That is
$$\frac{Q_1}{Q_2} = \frac{n_1}{n_2} \tag{6.1}$$

or
$$Q_1 n_2 = Q_2 n_1 \tag{6.2}$$

$$\frac{p_1}{p_2} = \left(\frac{n_1}{n_2}\right)^2 \tag{6.3}$$

or
$$\left(\frac{p_1}{p_2}\right)^{\frac{1}{2}} = \frac{n_1}{n_2} \tag{6.4}$$

*By decision of the twelfth Conférence Générale des Poids et Mesures in October, 1964 the old definition of the litre ($1.000\,028$ dm^3) was rescinded and the word litre was reinstated as a special name for the cubic decimetre. Neither the word litre nor its symbol l should be used to express results of high precision. While the litre is not part of the SI it is recognized that its use may be continued for some time, but it is recommended that except in special circumstances it should be progressively abandoned in conformity with international recommendations and replaced by $10^{-3}m^3 = dm^3$. Since water flow rates in hot water systems are not precise values it is convenient in this text to use the litre.

161

from Eq (6.1) and Eq (6.4)

$$\left(\frac{p_1}{p_2}\right)^{\frac{1}{2}} = \frac{Q_1}{Q_2} \tag{6.5}$$

or

$$\frac{p_1}{p_2} = \left(\frac{Q_1}{Q_2}\right)^2 \tag{6.6}$$

$$\frac{P_1}{P_2} = \left(\frac{n_1}{n_2}\right)^3 = \left(\frac{Q_1}{Q_2}\right)^3 \tag{6.7}$$

and by appropriate substitution

$$\frac{P_1}{P_2} = \left(\frac{p_1}{p_2}\right)^{\frac{3}{2}} \tag{6.8}$$

where Q = volume flow
       p = pump pressure
       n = pump speed
       P = power

If p = pump pressure
     η_p = pump efficiency as a decimal
     η_m = motor efficiency as a decimal

Then   Water power = $Q p$ (6.9)

$$\text{Pump power} = \frac{Q p}{\eta_p} \tag{6.10}$$

$$\text{Total power requirement} = \frac{Q p}{\eta_p \, \eta_m} \tag{6.11}$$

It should be noted that the water power is zero at no flow and generally increases with increased flow, even though the pressure is less. A pump operating at more than the design flow rate will therefore require more power, which may result in overloading the motor.

Performance data is most commonly shown by means of curves, Fig. 6.1, which relate pump capacity, pressure, power and efficiency for a constant pump speed.

The pressure/capacity curve may be either steep or flat, as shown in Fig. 6.2. Steep-curve pumps have the advantage that a change in the circuit resistance, due to scale deposits, results in only a small variation in capacity. Also, since the frictional resistance in a hot-water circulation cannot be determined very accurately, steep curve pumps may be preferred. Flat curve pumps are, however, also suitable, since only small variations in pressure result in large changes in pump capacity. This is a useful asset when balancing a number of circuits in a system. The choice of a pump will therefore depend upon the particular requirements of the circuit.

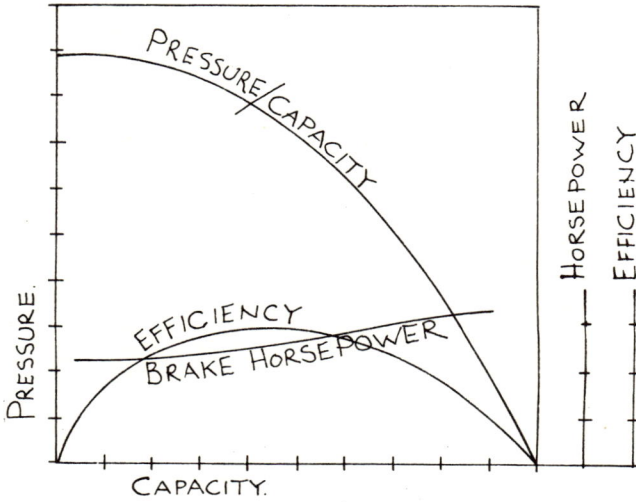

Fig. 6.1. Typical pump performance curves.

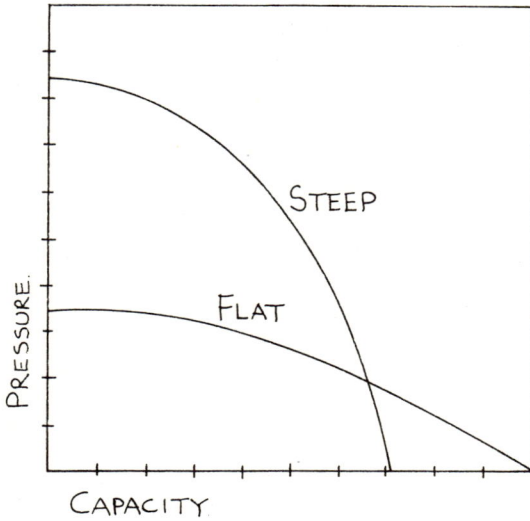

Fig. 6.2. Steep and flat head/capacity curves.

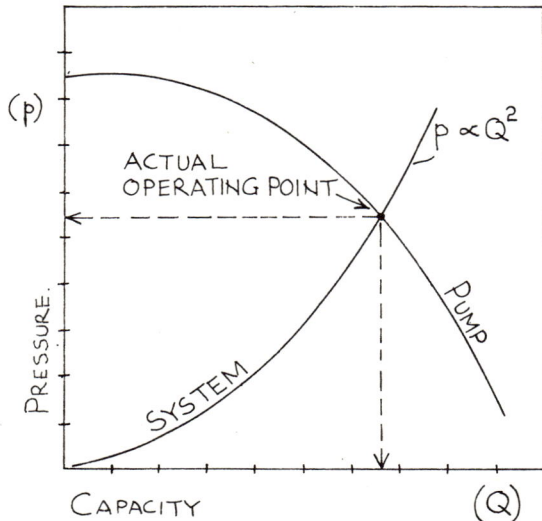

Fig. 6.3. Actual point of operation.

The point at which a pump will operate on its pressure/capacity curve is determined by the characteristic of the circuit to which it is connected. The resistance of the circuit may be taken with sufficient accuracy as varying with the flow rate approximately in proportion to the square of the change in velocity of the water or, alternatively, for constant cross-section area:

$$\frac{p_1}{p_2} = \left(\frac{Q_1}{Q_2}\right)^2 \qquad (6.12)$$

where p_1 = loss of total pressure at flow Q_1
         p_2 = loss of total pressure at flow Q_2

The intersection of the system characteristic curve, Eq (6.12), and the pump pressure/capacity curve indicates the actual operating conditions, Fig. 6.3. This follows, since, according to the first law of thermodynamics, the energy put into the water by the pump must be equal to the energy lost by the water as it overcomes the resistance offered to flow by the piping circuit.

The water pressure at any particular point in a closed circuit with the pump running is established by the algebraic sum of the static and residual pump pressure. Whether the pump pressure is additive or subtractive is determined by the relative pump and cold feed location. Consider the simple arrangement shown in Fig. 6.4. When the pump is started water will be drawn from the tank and passed into the vent pipe until the pressure due to the difference in water levels is equal to the residual pump pressure at "*b*".

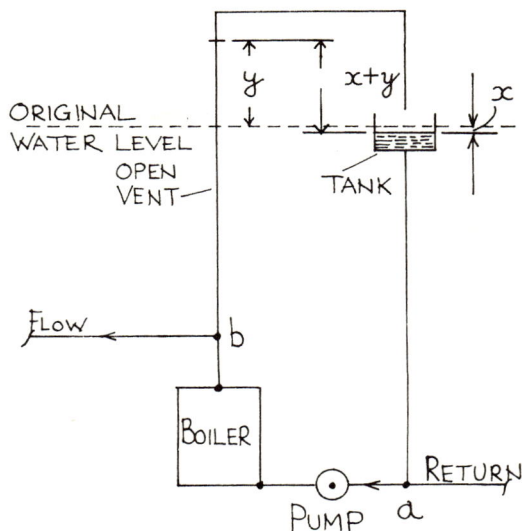

Fig. 6.4. Relative pump and cold feed position.

If p = pump pressure and p_{a-b} loss of pressure between "a" and "b" the difference in level between the water in the tank and the vent will be equivalent to the residual pump pressure $p - p_{a-b}$.

Let A_1 = cross-section area of tank
A_2 = cross-section area of vent pipe
x = change in water level in tank
y = change in water level in vent pipe
p_r = residual pump pressure, i.e. $p - p_{a-b}$
ρ = water density
g = standard gravity

Then
$$x \cdot A_1 = y \cdot A_2$$

and
$$p_r = \rho g x \left(\frac{A_1}{A_2} + 1\right) \tag{6.13}$$

For a tank 1 m × 1 m and a vent pipe 20 mm diam.:
$$A_1 = 1 \, \text{m}^2 \quad \text{and} \quad A_2 = 0.0003142 \, \text{m}^2$$

Assuming a residual pump pressure of 30 kN/m² and a water density of 1 000 kg/m³ and applying Eq (6.13):

$$30 = 1\,000 \times 9.81x \left(\frac{1}{0.0003142} + 1 \right)$$

from which $x = 0.001$ mm

The change in water level in the tank is found to be negligible, and it is apparent that a larger-diameter vent pipe would also have a negligible effect. For all practical purposes the point of entry of the cold feed into the system may be taken as the neutral point, i.e. the point at which the pressure remains constant whether the pump is running or not.

Since the water level in the tank is considered to remain unaltered, the distribution of pump pressure may be related to the position of the cold feed connection, as shown for four different arrangements in Fig. 6.5 and Fig. 6.6. In each case positive and negative signs are used to indicate an increase or decrease respectively in the static pressure due to the effect of the pump pressure.

The main features of each arrangement are given below:

Arrangement A. Most of the system operates at higher pressures when the pump is running than when the pump is stopped. This may be an advantage as far as air venting the heating apparatus is concerned. Also the higher pressure increases the anti-cavitation margin. On the other hand, the increase in pressure in the boiler may be a disadvantage. The main disadvantage is that the open vent pipe must be taken to a height well above the level of the tank to prevent discharge of water. This may not be possible in some tank rooms. The vent should not be taken outside the roof for fear of freezing the water in it.

Arrangement B. With this arrangement most of the system operates at pressures lower than static. Care should therefore be taken to ensure that high-level apparatus does not operate at sub-atmospheric pressures and hence draw air into the system through valve glands. The reduction in pressure in the boiler will reduce the anti-cavitation margin, and this should therefore be checked. The only advantage is that the vent pipe need not be extended above the level of the tank.

Arrangement C. This has the advantages of arrangement "A" but without the need to carry the vent above the feed tank. There is a reduction in boiler pressure.

Arrangement D. As for arrangement "C" but vent pipe must be taken above level of the tank.

For most installations the best position for the pump is in the flow main, with the open vent and cold feed on its suction side as shown at "C" in Fig. 6.6.

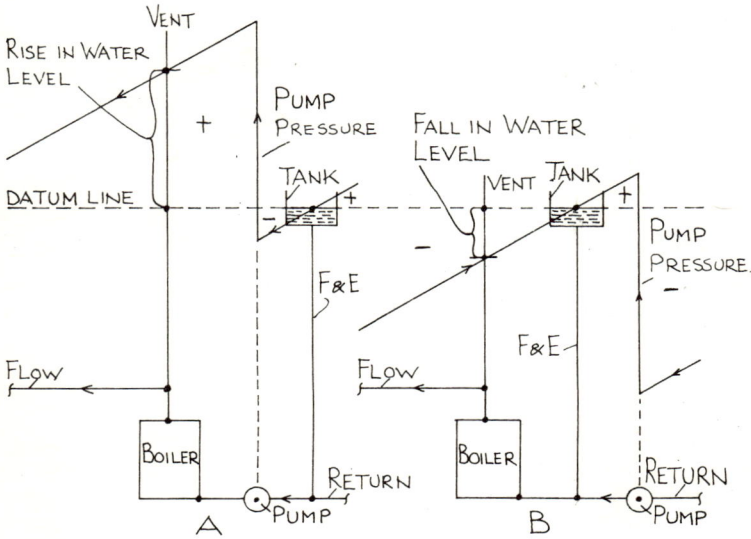

Fig. 6.5. Pump in return main.

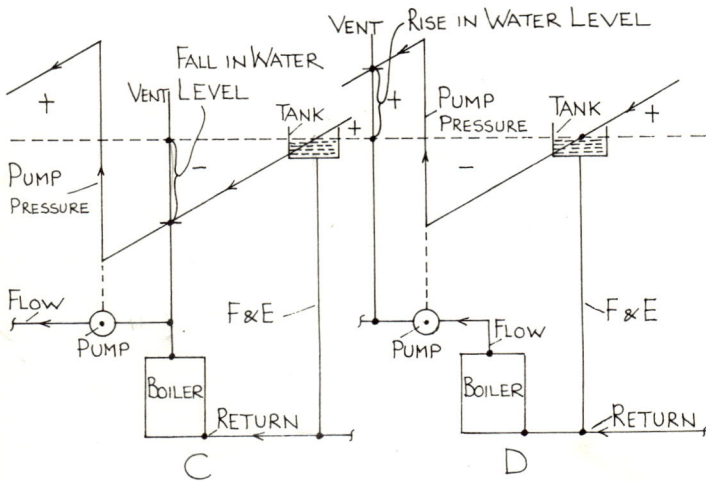

Fig. 6.6. Pump in flow main.

When a number of separate circuits, each requiring individual temperature control, are supplied from a common boiler plant interaction between
the circuits is inevitable. This may be reduced to a minimum if each circuit
has its own circulating pump rather than providing a single pump for the
combined loads. Such an arrangement is shown in Fig. 6.7, in which the
primary circuit merely supplies hot water to the common pipes A, B and C. In
all such cases it is essential that these pipes have a negligible pressure drop.

In the case of forced circulation hot-water supply systems the pump may
be installed in the flow or in the return, as shown in Fig. 6.8.

Example 6.1. During a test a pump running at 12 rev/s is found to have
the following characteristic:

p, kN/m^2	0	20	35	46	52	56
Q, l/s	3.5	3.1	2.3	1.5	0.75	0

(*a*) Draw the head/capacity curves for pump speeds of 12, 16 and
24 rev/s.

(*b*) If the pump is used in a system having a frictional resistance of
30 kN/m^2 for a water flow rate of 2.4 l/s what will be the actual operating
conditions at each pump speed.

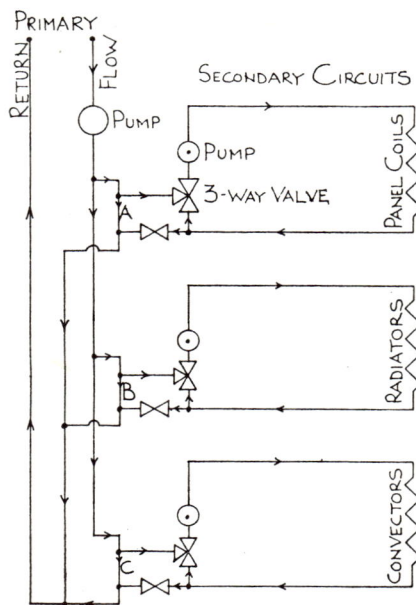

Fig. 6.7. Individually pumped primary and secondary circuits.

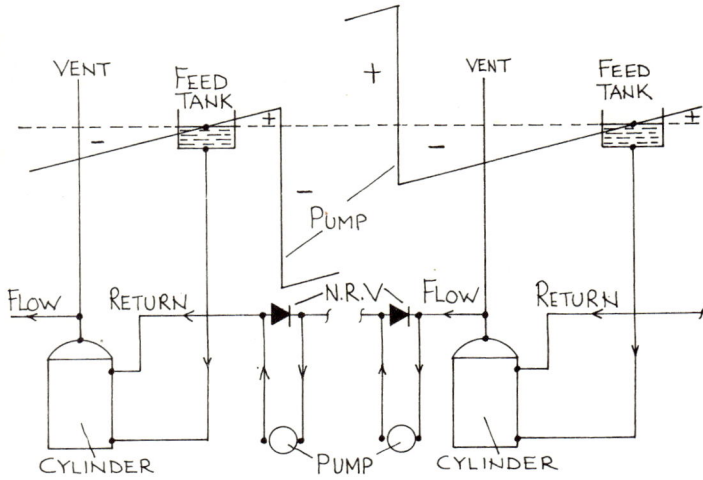

Fig. 6.8. Hot-water supply with forced secondary circulation.

(*a*) For a pump speed of 16 rev/s and using Eq (6.1)

$$\frac{Q_1}{Q_2} = \frac{12}{16}$$

i.e. $Q_2 = 1.33\, Q_1$

and from Eq (6.3)

$$\frac{p_1}{p_2} = \left(\frac{12}{16}\right)^2$$

i.e. $p_2 = 1.78\, p_1$

For a pump speed of 24 rev/s and proceeding as above

$$Q_2 = 2Q_1$$

and $p_2 = 4p_1$

Applying these correction factors to the given pump characteristic, the following data are obtained and plotted in Fig. 6.9.

12 rev/s		16 rev/s		24 rev/s	
p	Q	p	Q	p	Q
0	3.5	0	4.6	0	7.0
20	3.1	36	4.1	80	6.2
35	2.3	62	3.1	140	4.6
46	1.5	82	2.0	184	3.0
52	0.75	93	1.0	208	1.5
56	0	100	0	224	0

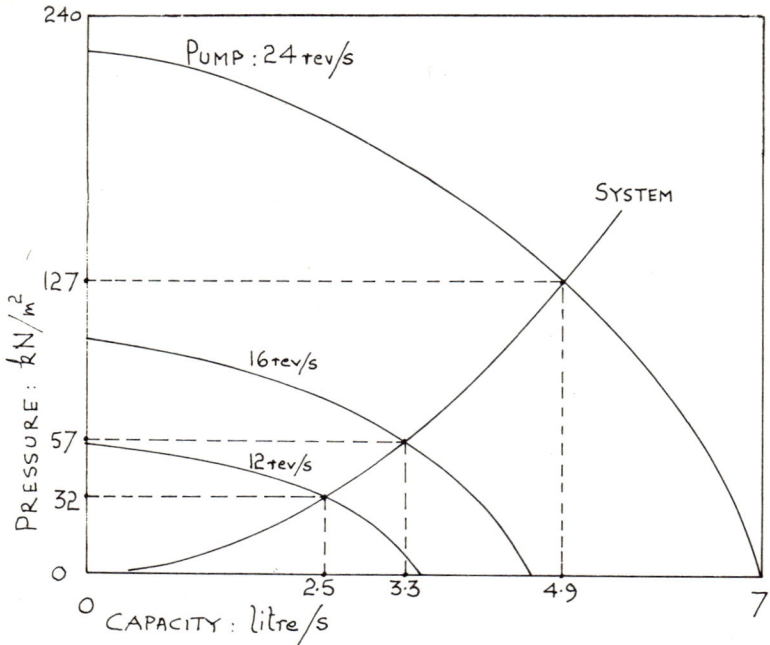

Fig. 6.9. Operating conditions for three different speeds.

(*b*) The system characteristic may be obtained from the given conditions by applying Eq (6.12) as follows:

$$p_2 = p_1\left(\frac{Q_2}{Q_1}\right)^2$$

i.e.   for half flow $p_2 = p_1(\tfrac{1}{2})^2 = 0.25p_1$
for 80 per cent flow $p_2 = p_1(0.8)^2 = 0.64p_1$

similarly:

p, kN/m^2	1.9	7.5	16.9	30	67.5	120
Q, l/s	0.6	1.2	1.8	2.4	3.6	4.8

The intersection of this curve, see Fig. 6.9, with the pump characteristic gives the following actual operating conditions:

Pump speed, rev/s	12	16	24
p, kN/m^2	32	57	127
Q, l/s	2.5	3.3	4.9

Example 6.2. A pump running at 16 rev/s has the following characteristics:

p, kN/m²	0	30	58	78	92
Q, l/s	4	3.5	2.5	1.5	0

Discuss the behaviour of two such pumps when connected in series and alternatively in parallel.

With two identical pumps connected in series the head developed is twice that developed by one of them, while the quantity of water delivered remains the same as for a single pump. The pressure/capacity characteristic for the series arrangement will therefore be:

p, kN/m²	0	60	116	156	184
Q, l/s	4	3.5	2.5	1.5	0

With the pumps running in parallel the opposite result is obtained, i.e. the quantity of water delivered by the two pumps is twice the quantity delivered by one of them, while the pressure developed remains the same as for a single pump. The characteristic for the parallel arrangement will therefore be:

p, kN/m²	0	30	58	78	92
Q, l/s	8	7	5	3	0

In each case the total power absorbed will be twice that of one pump.

The above pressure/capacity relationships are shown in Fig. 6.10, which also shows the single-pump characteristic and two system characteristics, curves A and B.

With system A it is seen that while the two pumps in parallel would give a greater flow, A_2, than a single pump, A_3, the series arrangement would give the highest flow, A_1. With system B, however, the parallel arrangement would give the highest flow, B_1, with the series flow, B_2, greater than the single-pump flow, B_3.

It is clear from this that two identical pumps connected in series may under certain conditions give a greater increase in flow rate than the same two pumps run in parallel. Thus, careful consideration should be given to the particular system characteristic before deciding to operate pumps in series or parallel to give an increased flow. Also with system A the series arrangement gives the greatest increase in pressure, while with system B the parallel arrangement gives the greatest increase.

When pumps are used in series they may be arranged as shown in Fig. 6.11.

Arrangement (*a*) is suitable when both pumps are required to operate simultaneously. If pump A is started first the water inertia will "turbine" the impeller of pump B, and so assist starting; the opposite effect occurs with pumps in parallel unless non-return valves are used as in Fig. 6.13.

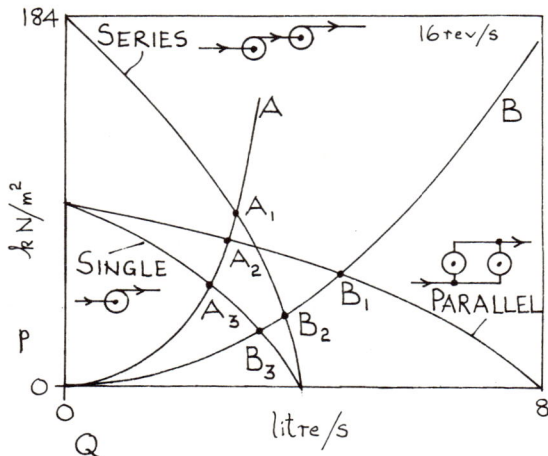

Fig. 6.10. Analysis of series and parallel arrangements.

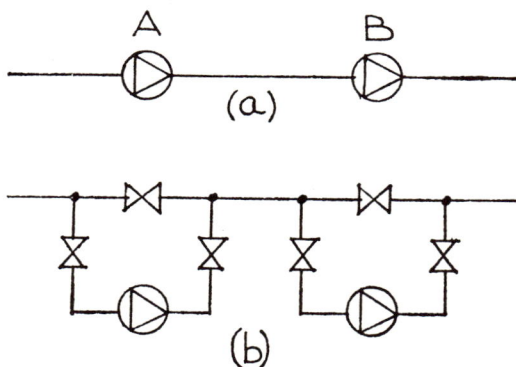

Fig. 6.11. Alternative series arrangements.

With only one of the two pumps operating, the idle pump offers an extra but generally negligible resistance to flow. Arrangement (*b*) is preferable in cases where either one or both pumps may be required to operate to meet a variable load. Either pump may be used as a stand-by and may be isolated from the main circulation for service and repair. Consider Fig. 6.12, which shows a series arrangement of two identical pumps. Point A indicates the combined effort of both pumps with each pump operating at condition B whereas point C gives the conditions for a single pump operating alone. It

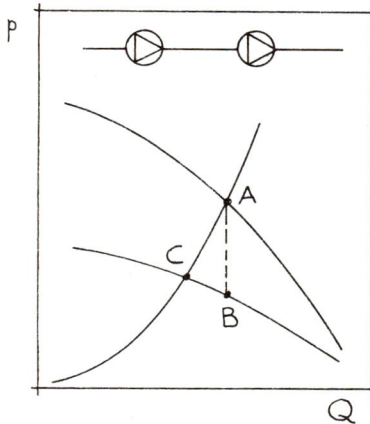

Fig. 6.12. Series arrangement.

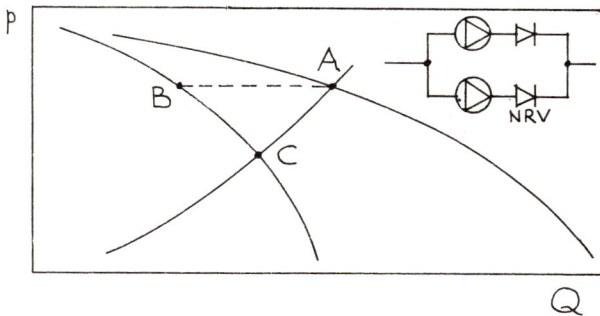

Fig. 6.13. Parallel arrangement.

should be noted that during combined operation the power requirement of each pump is fixed by condition B and that the power requirement of a single pump, point C, is less. There is generally a power reduction when a single pump of a series arrangement is used and a power increase when a single pump of a parallel arrangement is used. This should be taken into account when assessing the annual running costs of alternative pumping arrangements, particularly for combined winter heating and summer cooling systems.

Two identical pumps are shown operating in parallel in Fig. 6.13.

Point A indicates the combined effect of both pumps with each running at condition B, and point C gives the condition for a single pump operating alone. When both pumps are running the power requirement of each will be less than for a single pump.

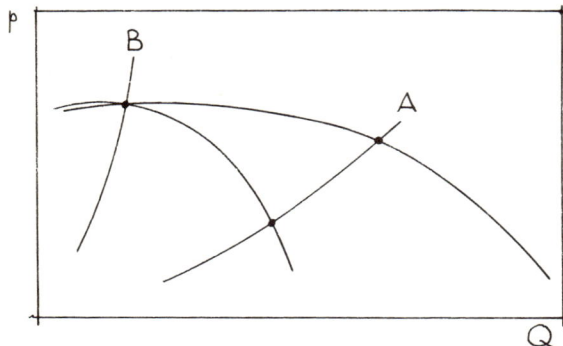

Fig. 6.14. Humped characteristics in parallel.

Pumps having "humped" characteristic curves are only suitable for use in parallel if the operating point is well away from the humped portion of the curves, as with system A in Fig. 6.14. There is no advantage to be gained by using these pumps in parallel for system B.

The use of pumps in series or parallel frequently enables smaller "in-line" type pump units to be used, thus saving the cost of pump and motor bases. With very large systems requiring a wide range of flow rates the use of pumps in series, parallel or combined series – parallel arrangements may result in minimum power requirements and running costs. Fig. 6.15 shows the range of operating conditions (A, B, C and D) for a typical series – parallel arrangement of four identical pumps.

Example 6.3. The hot-water system shown in Fig. 6.16 consists of two circuits A and B arranged in parallel and supplied from a common circuit C having a resistance of 15 kN/m^2 for a flow rate of 9 l/s. Circuits A and B each have a resistance of 9 kN/m^2 for a flow rate of 5 l/s, and each includes a pump having the following characteristic:

p, kN/m^2	20	16	12
Q, l/s	2	4	5

Determine the quantity of water flowing in each circuit and the actual pressure at which the pumps operate.

Considering the parallel arrangement and using Eq (6.12), the characteristic of either circuit A or B, taken at the same flow rates as the given pump characteristic, will be

p, kN/m^2	1.44	5.76	9.0
Q, l/s	2	4	5

At each of the above flow rates the residual pump pressure available for the common circuit C will be: $20 - 1.44 = 18.56$, $16 - 5.76 = 10.24$ and $12 - 9 = 3.0$. Since the pumps are operating in parallel circuits, the flow

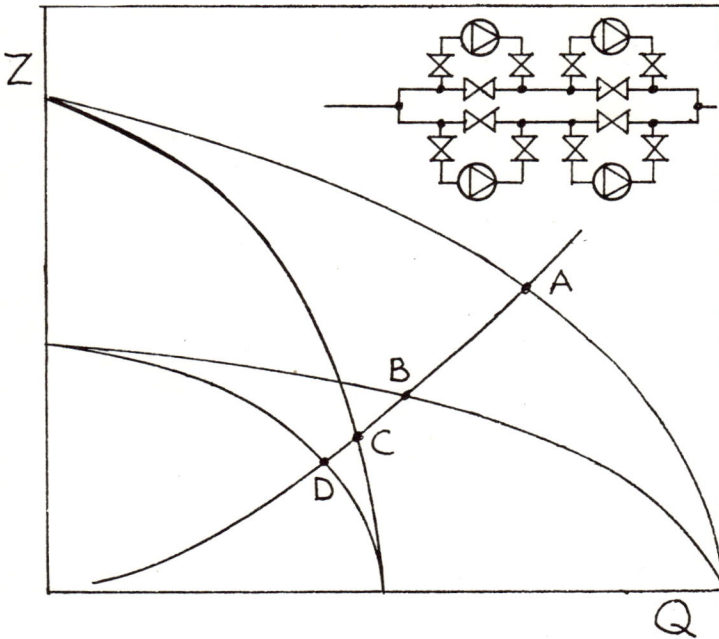

Fig. 6.15. Combined series – parallel arrangement.

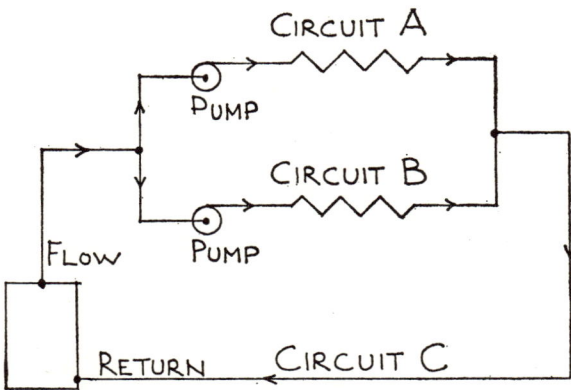

Fig. 6.16. Parallel pumped circuits.

rates will be twice the flow rates of a single pump. The residual pressure/ capacity characteristic will therefore be:

$$
\begin{array}{llll}
\text{Residual pressure, kN/m}^2 & 18.56 & 10.24 & 3.0 \\
2Q, \text{l/s} \dots\dots\dots\dots & 4 & 8 & 10
\end{array} \quad \dots\dots \quad (1)
$$

and from Eq (6.12) the characteristic of the common circuit C will be:

$$
\begin{array}{llll}
p, \text{kN/m}^2 \dots\dots & 3.75 & 6.7 & 15 \\
Q, \text{l/s} \dots\dots\dots & 4.5 & 6 & 9
\end{array} \quad \dots\dots \quad (2)
$$

The characteristics 1 and 2 are plotted in Fig. 6.17, and from their intersection the flow rate in the common circuit 3 is found to be 7.8 l/s for a residual pump pressure of 11 kN/m². The flow rate in each circuit A and B will therefore be 7.8 ÷ 2 = 3.9 l/s, for which, using Eq (6.12), the resistance will be $9(3.9/5.0)^2 = 5.47$ kN/m². The pressure at which each pump will operate will then be $11 - 5.47 = 5.53$ kN/m².

Example 6.4. Consider the piping arrangement shown in Fig. 6.18, and using the data listed below, determine the quantity of water flowing in each circuit and the pressure at which each pump will operate.

Circuit	A	B	C
p, kN/m²	6	16	30
Q, l/s	6	6	10

Characteristic of each pump:

p, kN/m²	35	26	17
Q, l/s	2.5	4.5	6

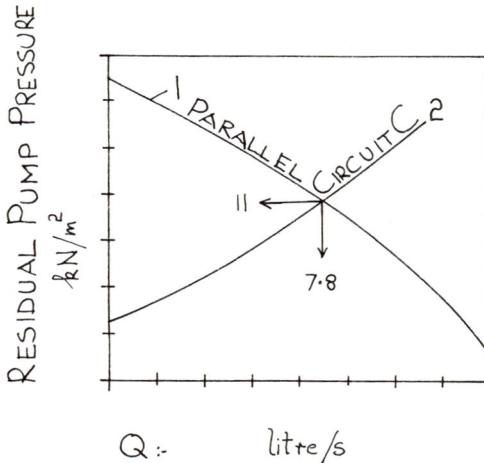

Fig. 6.17. Residual head/capacity curve for pumps in parallel.

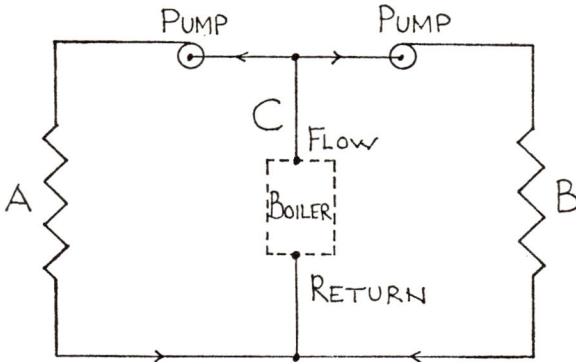

Fig. 6.18. Parallel circuits across a common boiler.

Consider the parallel circuits A and B and using Eq (6.12), make up the following table, which also shows the residual pump pressure in each circuit.

Q	Pump	Circuit A		Circuit B	
	p	p_A	$p-p_A$	p_B	$p-p_B$
2.5	35	1.04	34	2.8	32.2
4.5	26	3.38	22.6	9	17
6	17	6	11	16	1

The characteristic of circuit C is, using Eq (6.12):

$p, kN/m^2$	30	19.2	7.5
$Q, l/s$	10	8	5

The above data is plotted in Fig. 6.19 in which:

Curve A = residual pump pressure/capacity characteristic of circuit A
Curve B = residual pump pressure/capacity characteristic of circuit B
Curve C = pressure/capacity characteristic of circuit C
Curve D = combination of curves A and B in parallel.

From the intersection of curves C and D the flow rate in circuit C is found to be 8.5 l/s for a residual pump pressure of 22 kN/m². The corresponding flow rates in circuits A and B are 4.6 and 3.9 l/s respectively. Then, using Eq (6.12):

For circuit A

$$p_A = 6\left(\frac{4.6}{6}\right)^2 = 3.54 \text{ kN/m}^2$$

Then, since

$$p - p_A = 22 \text{ kN/m}^2$$

$$p = 22 + 3.54 = 27.54 \text{ kN/m}^2$$

Similarly, for circuit B:

$$p_B = 16\left(\frac{3.9}{6}\right)^2 = 6.76 \text{ kN/m}^2$$

$$p = 22 + 6.76 = 28.76 \text{ kN/m}^2$$

Summary of actual operating conditions:

Circuit	A	B	C
p, kN/m^2	27.54	28.76	22
Q, l/s	4.6	3.9	8.5

Example 6.5. A water circulation system is shown in Fig. 6.20. Using the data given below determine the actual pump pressure capacity and the flow rate in each of the circuits A and B.

Data:

Circuit	A	B	C
p, kN/m^2	30	35	43
Q, l/s	7.5	6	11.5

Pump characteristic:

p, kN/m^2	65	50	30
Q, l/s	7	10	12

In general:

$$p \propto Q^2$$

i.e.

$$p = kQ^2$$

Then, since the total resistance of each of the parallel circuits A and B must be the same and because $Q_C = Q_A + Q_B$

$$\frac{1}{\sqrt{k_{A-B}}} = \frac{1}{\sqrt{k_A}} + \frac{1}{\sqrt{k_B}}$$

$$= \frac{7.5}{\sqrt{30}} + \frac{6.0}{\sqrt{35}} = 2.38$$

The parallel arrangement A and B is in series with circuit C for which

$$k_C = \frac{43}{11.5^2}$$

and for the combined system

$$k = \frac{43}{11.5^2} + \frac{1}{2.38^2} = 0.5$$

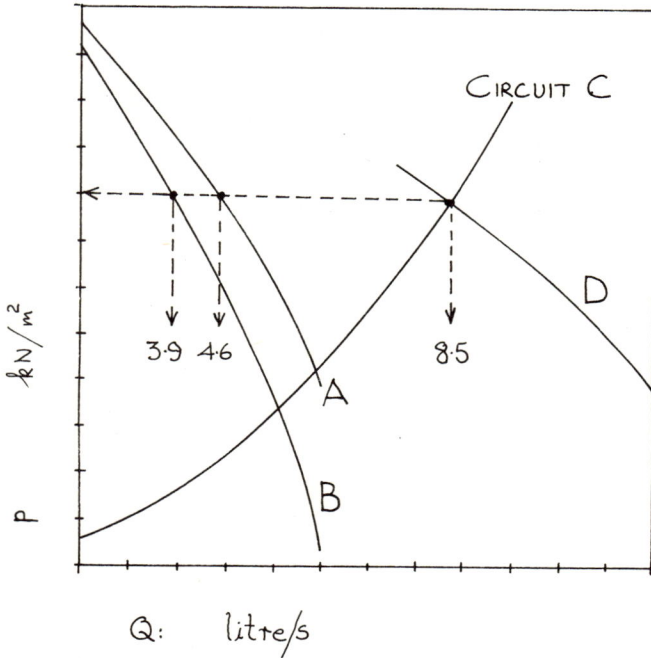

Fig. 6.19. Characteristic curves for parallel pumps and common circuit.

Thus for a flow of 11.5 l/s

$$p = 11.5^2 \times 0.5 = 66 \text{ kN/m}^2$$

and the characteristic of the combined system will therefore be:

| p, kN/m^2...... | 66 | 42.3 | 16.5 |
| Q, l/s .... | 11.5 | 9.2 | 5.75 |

This is plotted on the pump characteristic in Fig. 6.21, and from the intersection the actual operating conditions of the pump are found to be 50 kN/m^2 and 10 l/s. Then:

$$p_C = 43\left(\frac{10}{11.5}\right)^2 = 32.5 \text{ kN/m}^2$$

The balance of $50 - 32.5 = 17.5$ kN/m^2 is common to the parallel circuits A and B, and the flow rates will therefore be:

$$Q_A = 7.5\sqrt{\frac{17.5}{30}} = 5.75 \text{ l/s}$$

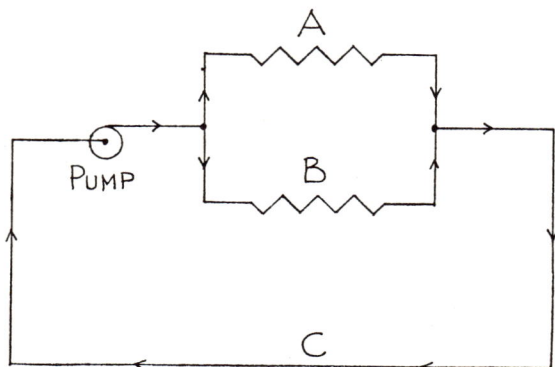

Fig. 6.20. Parallel circuits with a single pump.

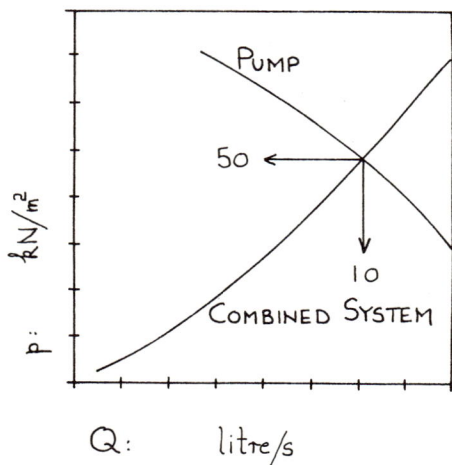

Q: litre/s

Fig. 6.21. Characteristic curves for combined system and pump.

and $$Q_B = 6.0 \sqrt{\frac{17.5}{35}} = 4.25 \text{ l/s}$$

or $$Q_B = 10 - 5.75 = 4.25 \text{ l/s}$$

Example 6.6. A hot-water system has an estimated flow rate of 6 l/s and a frictional resistance of 20 kN/m². The pump selected is belt driven and has the following characteristic when running at 16 rev/s.

p, kN/m²	17	20	23
Q, l/s	7.5	6.5	5.0

(*a*) Determine the actual operating conditions for the system when the pump speed is 16 rev/s.

(*b*) Determine the pump speed required to meet the estimated duty.

(*a*) Using Eq (6.12), the system characteristic will be

p, kN/m^2	27.2	20	13.9	8.9
Q, l/s	7	6	5	4

This is plotted in Fig. 6.22, and from the intersection with the pump characteristic the actual operating conditions for the system are found to be 21 kN/m^2 and 6.15 l/s. Then from Eq (6.1),

$$n_2 = \frac{16 \times 6}{6.15} = 15.6 \text{ rev/s}$$

Example 6.7. Two circuits, each with its own circulating pump, are connected to a boiler as shown diagrammatically in Fig. 6.23. Using the data given below, determine the position of the neutral point in the right-hand circuit.

Data:

L.H. Circuit		R.H. Circuit	
Pipe No.	Resistance, kN/m^2	Pipe No.	Resistance, kN/m^2
1	4	7	16
2	6	8	20
3	15	9	12
4	20	10	12
5	20	11	20
6	5	12	12
		13	4
Total:	70	Total:	96

The position of the neutral point in the right-hand circuit is determined by the position of the feed and expansion pipe connection in the left-hand circuit. Fig. 6.24 shows the change in resistance for both circuits and indicates that the neutral point of the right-hand circuit will be at "X".

Problems

1. A hot-water heating circuit has an estimated flow rate of 6.82 l/s and a resistance of 28.9 kN/m^2. The pump selected has the following characteristic:

Flow rate, l/s	5.3	6.1	6.8	7.6
Pressure, kN/m^2......	31.8	28.9	26.6	22.8

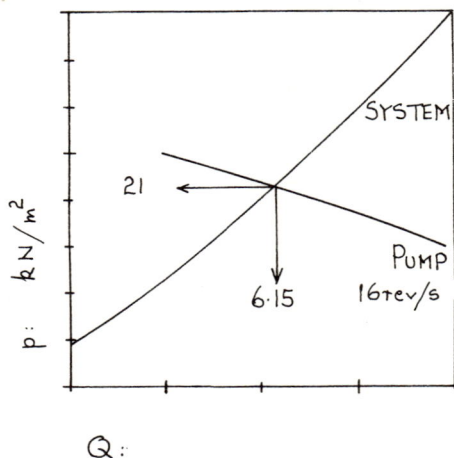

Fig. 6.22. Actual operating conditions, single pump and system.

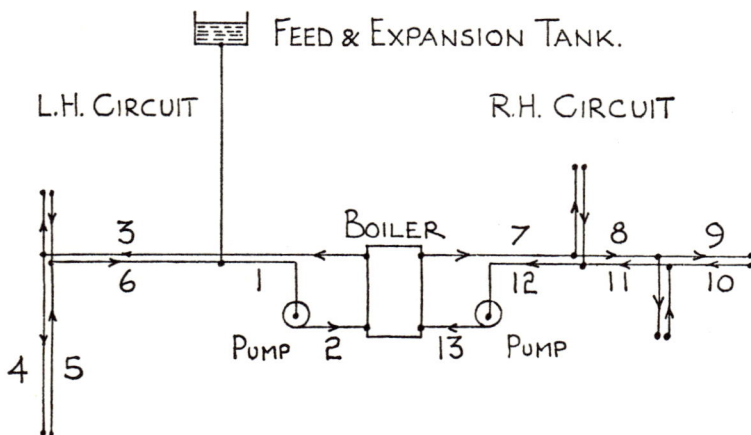

Fig. 6.23. Two systems with common feed and expansion tank.

What will be the actual operating conditions?

Ans.: 6.67 l/s; 26.9 kN/m².

2. A forced-circulation hot-water heating system has an estimated flow rate of 7.6 l/s and a resistance of 20 kN/m². The pump selected is belt driven and has the following characteristic when running at 20 rev/s.

Q, l/s	7.6	5.9	4.8	3.3
p, kN/m²	17.3	21.7	23.8	26.2

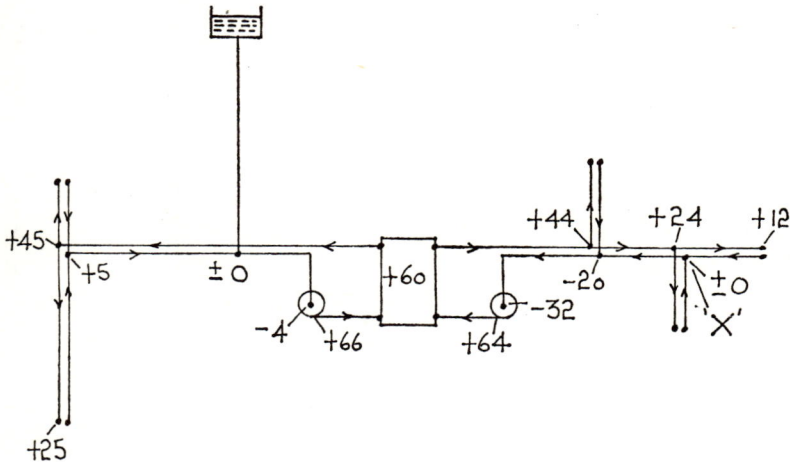

Fig. 6.24. Determination of neutral point.

(i) Determine the actual operating conditions for the system when the pump speed is 20 rev/s.

(ii) Determine the pump speed required to meet the estimated duty for the system.

Ans.: (i) 7.2 l/s; 18.2 kN/m²; (ii) 21.1 rev/s.

3. A pump has the following characteristic:

p, kN/m²	2.9	5.8	8.7	11.6
Q, l/s	3.25	2.7	2.0	1.2

What will be the actual operating conditions when used in a system having an estimated flow rate of 2.27 l/s and a resistance of 5.78 kN/m² when:

(i) a single pump is used;

(ii) two pumps are used in parallel;

(iii) two pumps are used in series?

Ans.: (i) 2.42 l/s, 4 kN/m²; (ii) 3 l/s, 10.4 kN/m²; (iii) 2.8 l/s, 11 kN/m².

4. Two circuits A and B are arranged in parallel and supplied from a common circuit C. Using the data listed below, determine the quantity of water flowing in each circuit and the actual head at which the pumps operate.

Data:

Circuit A and circuit B:

Resistance 5.78 kN/m² when passing 3 l/s

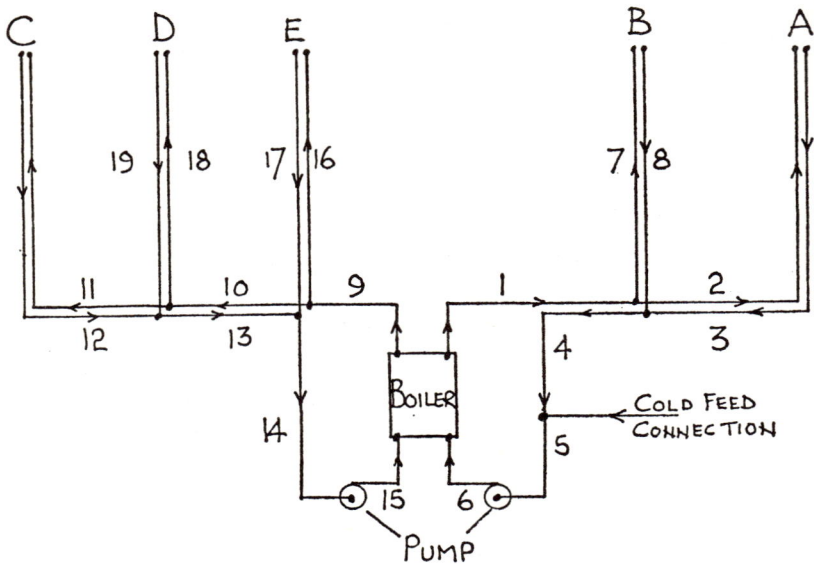

Fig. 6.25. Two circuits connected to a common boiler.

Circuit C:

Resistance 10.1 kN/m² when passing 5.68 l/s

Characteristic of pumps in circuits A and B:

p, kN/m²	11.6	10.4	8.95	6.95
Q, l/s	0.75	1.5	2.3	3.0

Ans.: 9.2 kN/m² and 2.2 l/s.

5. A certain pump has the following characteristics:

Q, l/s	0	3.8	7.6	11.4	15.2	18.9
p, kN/m²	111	113	114	108	98	77
Pump power, kW	0.8	1.1	1.5	1.8	2.2	2.4
Efficiency, per cent ..	0	40	62	71	71	64

Determine the actual operating conditions when used in the following systems:

(*a*) System A: 15 l/s and 116 kN/m²
(*b*) System B: 15 l/s and 87 ” ”
(*c*) System C: 15 l/s and 72 ” ”
(*d*) System D: 11.4 l/s and 116 kN/m²

Ans.: (*a*) 14.2 l/s; 101.2 kN/m²; 2.16 kW; 72 per cent
    (*b*) 15.9 " ; 95.4 "   " ; 2.24 " ; 70 " "
    (*c*) 17.2 " ;· 89.6 "   " ; 2.38 " ; 68 " "
    (*d*) 11.1 " ; 108.4 "   " ; 1.87 " ; 70 " "

6. Two circuits connected to a common boiler are shown in Fig. 6.25.

(*a*) Determine the diameter of each pipe. Assume a water flow velocity of about 1 m/s.

(*b*) Determine the duty of each pump.

(*c*) Determine the position of the neutral point in the left-hand circuit.

Data:

Section No.	Length, m		Unit	Water flow rate, kg/s	Pressure drop, kN/m²
1	15.2		A	1.26	11.56
2	45.7		B	1.01	11.56
3	45.7		C	1.13	11.56
4	9.1		D	0.19	11.56
5	9.1		E	1.01	11.56
6	3.7				
7	30.5				
8	30.5		1. Neglect pressure drop across boiler.		
9	61.0		2. Neglect static pressure changes.		
10	30.5		3. Allow 30 per cent extra on given lengths		
11	45.7		to allow for fittings resistance.		
12	45.7				
13	30.5				
14	61.0				
15	6.1				
16	30.5				
17	30.5				
18	45.7				
19	45.7				

Ans.:

(*a*) Section ...	1	2	3	4	5	6	7	8	9	10
Diam., mm	65	50	50	65	65	65	40	40	65	50
Section ...	11	12	13	14	15	16	17	18	19	
Diam., mm	40	40	50	65	65	40	40	32	32	

(*b*) Right-hand pump: 2.27 l/s and 28.9 kN/m²,
    Left-hand pump: 2.95 l/s and 75.1 kN/m².

(*c*) The neutral points in the left-hand circuit are in sections 11 and 18.

7: Steam Systems

Introduction

Steam is used for many industrial processes that require constant-temperature conditions and is also suitable as a primary medium for distributing heat in hospital and factory centres. For heating purposes it is usual to use only the latent heat in the steam, the heat in the residual condensate often being lost. Since the value of the latent heat is greatest at low pressure, steam for space heating should be used at the lowest pressure compatible with the pressure drop in the distribution network, condensate disposal and with the permissible heat output of individual heating surfaces.

Condensate is generally regarded as an undesirable by-product of steam heating, and wherever practicable it should be returned to the boiler hot-well for re-use as feed-water rather than passed to drain. This increases heat efficiency of the cycle and also keeps water-treatment problems to a minimum.

High-temperature processes require high-pressure steam and problems often arise over the disposal of the high-temperature condensate. Two possible solutions are: (a) to pass the condensate through a heat exchanger and discharging it at low temperature, and (b) to pass the condensate to a flash-steam recovery plant. When condensate at high pressure and saturation temperature is discharged to atmosphere or a region of lower pressure it will boil and form some steam at the lower pressure. The steam formed in this way is called flash steam.

Consider the case of 1 kg of condensate at saturation temperature at 0.7 MN/m^2 absolute pressure* when discharged to atmosphere. From steam tables it will be seen that the specific enthalpy of the condensate at these conditions in 697 kJ/kg, while that at atmospheric pressure is

---

* Pressure is force per unit area and since the unit of force is defined as that force which will impart unit acceleration to unit mass, the coherent unit for pressure in the International System of Units is $\dfrac{\text{kg m}}{\text{m}^2 \text{s}^2}$ or N/m^2. While units with special names such as the bar (1 bar = 10^5N/m^2) may continue to be used for some time, particularly in steam tables, it is recommended that they should be progressively abandoned in conformity with international recommendations. The SI unit and its recommended multiples e.g. MN/m^2 and kN/m^2 are therefore used in this book.

417 kJ/kg. There is therefore a surplus of $697 - 417 = 280$ kJ/kg, and as the latent heat of vaporization at atmospheric pressure is 2 258 kJ/kg $280 \div 2\ 258 = 0.12$ kg of steam will be formed at atmospheric pressure and $1 - 0.12 = 0.88$ kg of the initial condensate will remain as water, but now at atmospheric pressure and saturation temperature. The amount of heat which can be recovered in this way is comparatively small, and unless it can be used locally the transmission losses are likely to outweigh any other advantages, particularly since the collection of flash steam requires additional capital expenditure on plant.

Many industrial processes require steam at both high and low pressure. In such cases, and if copious quantities of high-pressure condensate are available, the possibility of using flash steam for the low-pressure plant should always be investigated as it may result in a fuel saving. At times there may be insufficient flash steam to meet the low-pressure requirements. When this occurs live steam is passed to the low-pressure plant via a steam-pressure reducing valve.

The theoretical arrangement of a flash steam system is shown diagrammatically in Fig. 7.1, in which, neglecting heat and pressure losses along the pipes:

x = live-steam make-up at p_2 from steam at p_1 and having a specific enthalpy of saturated vapour at p_1, kg/s

y = condensate discharged from the high-pressure plant and having a specific enthalpy of saturated liquid at p_1, kg/s

z = flash steam formed at p_2 and having a specific enthalpy of saturated vapour at p_2, kg/s

$x + y - z$ = residual condensate at p_2 and having a specific enthalpy of saturated liquid at p_2, kg/s

p = steam pressure, N/m^2

h_f = specific enthalpy of saturated liquid, J/kg

h_g = specific enthalpy of saturated vapour, J/kg

Suffixes 1 and 2 refer to the high- and low-pressure conditions respectively. A heat and mass balance for the system would then be:

$$x \cdot h_{g1} + y \cdot h_{f1} = z \cdot h_{g2} + (x + y - z)h_{f2} \qquad (7.1)$$

Eq (7.1) may be modified to include the quality of the steam and used to solve all types of flash-steam problem.

The method of using steam for heating purposes depends to some extent upon the size of the installation. Very small systems operate under totally enclosed conditions with the condensate returning to the boiler by natural circulation. Another form of closed system operates with steam at sub-atmospheric pressures and depends upon a vacuum pump for returning the

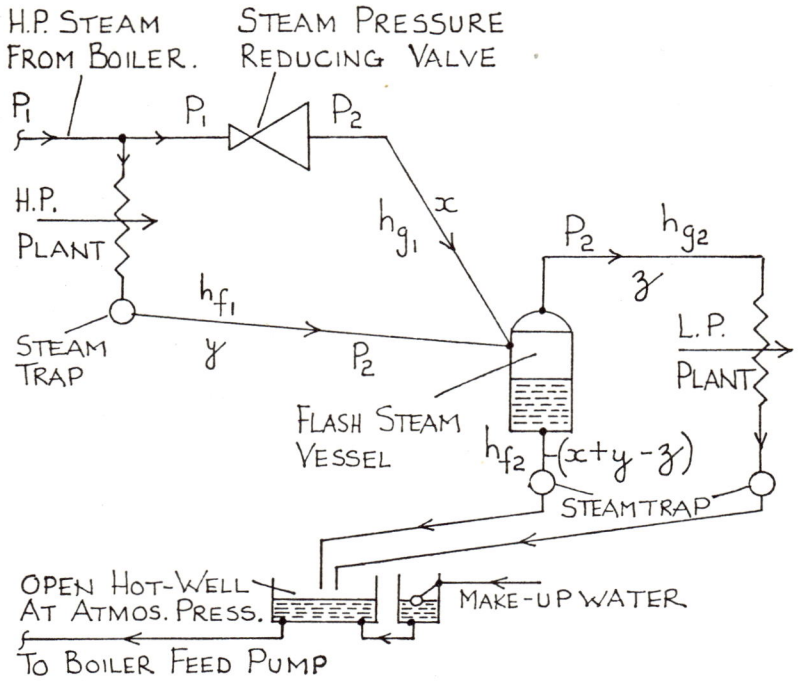

Fig. 7.1. Flash steam recovery and use, basic arrangement.

condensate to the boiler. This system is suitable for fairly large buildings and may be controlled to operate over a wide range of steam temperatures.

For large industrial space and process heating the open type of steam system is generally used. With this system, steam is passed from the boilers to a steam-pressure reducing station, from where it is distributed at the required pressures. A separate condensate return is generally run for each service provided. Since the hot-well is open to atmosphere, there is always a loss of water due to evaporation, and some make-up water is necessary. As a result, the hot-well temperature rarely exceeds $85°C$; if higher temperatures are required the feed water may be passed, after the pumps, through a feed-water heater situated in the flue outlet from the boilers.

In certain circumstances it may be possible to use the steam exhausted from non-condensing engines for space or process heating purposes. In such cases care must be taken to ensure that the permissible back pressure on the engine is not exceeded. This may be done by installing an exhaust relief valve in the main exhaust header. Exhaust steam is contaminated with engine lubricating oil, and must be passed through an oil separator before it is used in any heating equipment. A live-steam make-up via a steam-pressure reducing valve should be included in the design for use at times when there is insufficient exhaust steam to meet the secondary heating requirements. Provision must also be made for condensing the exhaust steam in the usual way at times when it is not required for space or process heating. The use of exhaust steam is conducive to fuel economy, and should always be considered.

With all forms of steam heating it is necessary to remove the condensate from the heating apparatus without losing live-steam in the process. Steam traps were developed to meet this requirement, and are now available in a variety of types that: operate at sub-atmospheric to high pressure, discharge large or small amounts of condensate, discharge condensate as soon as it is formed or after it has given up some of its heat, discharge condensate continuously or intermittently. They may be broadly classified as follows:

Mechanical float operated:
Open-top bucket
Inverted bucket
Spherical free float
Ball valve
Slide valve

Thermostatic:
Liquid expansion
Bi-metallic expansion
Tube expansion
Balanced pressure bellows

Other types:
Thermodynamic
Labyrinth
Lifting or pumping

Some of the above steam traps are fitted with automatic air and steam lock release valves. Unless a de-aerator is used, air will enter the system in solution in the boiler feed-water. It may also be drawn into the system through the packing glands of valves at times when the system is cooling down and tending to cause a partial vacuum. Air and other non-condensable gases in the steam may retard the free flow of condensate, create cold spots on the heating surface and reduce the total heat transfer. Dalton's law of partial pressures states that the total pressure exerted by a mixture of gas and vapour is equal to the sum of the pressures each would exert if occupying the space alone, i.e.

$$p_t = p_g + p_s$$

where p = pressure and subscripts t, g and s refer to the total pressure and partial pressures of the gas and vapour respectively.

Consider a steam system operating at a total pressure of $0.7\,MN/m^2$ abs.

and containing a steam-air mixture of 9 parts of steam to 1 part of air. The partial pressures of the steam and air will be:

$$\text{Steam:} \quad 0.7 \cdot \frac{9}{10} = 0.63 \, \text{MN/m}^2 \text{ abs}$$

$$\text{Air:} \quad \quad 0.7 \cdot \frac{1}{10} = 0.07 \, \text{MN/m}^2 \text{ abs}$$

Since the air will be at the same temperature as the steam, the maximum temperature in the system will correspond to the partial steam pressure of $0.63 \, \text{MN/m}^2$ abs. From steam tables this is found to be 161°C, which is seen to be 4°C below the temperature corresponding to the total pressure in the system. In addition to discharging condensate without allowing steam to pass, the steam trap should therefore be capable of passing air and any other non-condensable gases which may be present.

The following brief details are given to show only the main principles of steam-trap design and operation; full constructional details, together with capacity and pressure drop data, are readily available in manufacturers' literature.

Mechanical steam traps use density difference, and hence buoyancy effects, to differentiate between steam and condensate. Most float types discharge the condensate at the same rate as it flows into the trap and at practically the same temperature as that of the steam, and therefore produce some flash steam at the outlet. These traps are preferable in cases where it is necessary to remove large quantities of condensate quickly and fairly continuously. They will, however, operate intermittently on light loads. The open-top bucket trap is particularly suitable for intermittent operation, but should not be used in exposed positions, where it may freeze-up in winter.

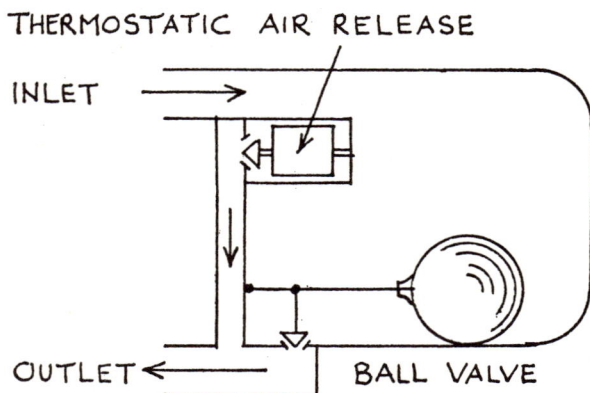

Fig. 7.2.

The most popular type for low and medium pressures is possibly the ball-valve float trap with thermostatic air release, shown diagrammatically in Fig. 7.2. Since the valve mechanism is always below water level, it is practically impossible for steam to blow straight through this trap. The capacity of the trap is a function of the pressure differential across it, the size and shape of the valve and seat and the amount of valve movement, which in turn is a function of the buoyancy of the ball float and the mechanical advantage of the lever system.

Thermostatic traps operate on changes in temperature of the thermostatic element and are usually pre-set by the manufacturer to close when the temperature of the condensate flowing through them rises to within about 5°C of the steam temperature. The liquid, bi-metallic and tube expansion types, however, may be readily adjusted to discharge condensate at any temperature up to steam temperature. When these traps are required to discharge direct to atmosphere the condensate must be held in the heating apparatus until it cools below 100°C. If this is not permissible a cooling-leg must be included between the heating apparatus and the trap, otherwise flooding of the apparatus may occur. This is particularly important with forced-air heater batteries and unit heaters because any condensate held back in the battery quickly cools by forced convection and causes undesirable temperature gradients in the discharge air stream. The principle of operation of expansion-type traps is shown in Fig. 7.3. (*a*) and (*b*).

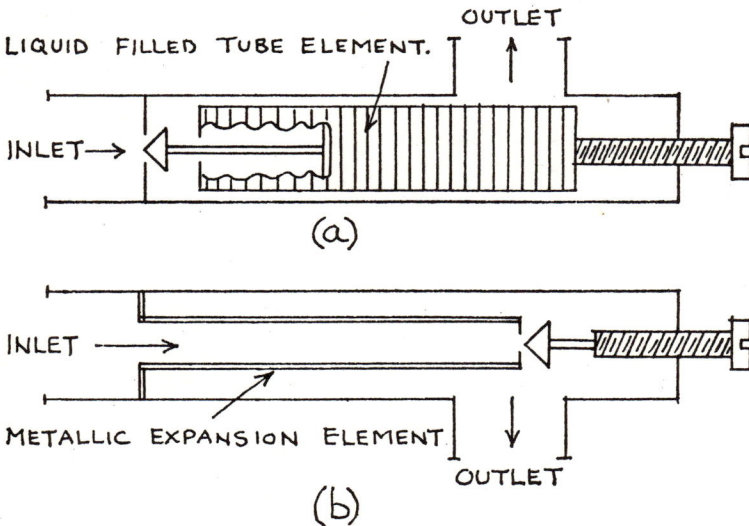

Fig. 7.3. Expansion type thermostatic traps.

Balanced-pressure thermostatic traps operate on change of phase, and hence change in pressure, of a fluid sealed within the motor element, which may be either a corrugated bellows or a diaphragm. The element is partly filled with a fluid which may be either water, a volatile fluid or a mixture of the two, depending on the particular type of element used and the required temperature − pressure relationship. During manufacture the element is sealed under a vacuum to ensure that the boiling point of the fluid is about 5°C lower than that of water. This pressure − temperature differential remains substantially constant over the working range of the trap. Consider such a trap, Fig. 7.4, draining condensate from a system operating at a pressure of 0·21 MN/m² abs, i.e. at 122°C. The pressure within the element will be equal to the external pressure when the condensate temperature is 117°C. As the temperature of the condensate in the trap body rises above 117°C the element absorbs heat and the fluid begins to boil, and when its vapour pressure exceeds 0·21 MN/m² abs the element will expand and seal the outlet. The trap re-opens as soon as the condensate within it cools to the condensation temperature of the fluid inside the element. These traps are not recommended in cases where superheated steam is used, because the vapour pressure within the element may become excessive and rupture the bellows or diaphragm A cooling leg may be required if this type of trap is used to drain air-heater batteries.

There are three types of thermodynamic trap: impulse, labrinth and kinetic energy. They depend on the thermodynamic properties of steam and hot condensate for their operation. When condensate at or near steam temperature is passed to a region of lower pressure it will boil and form some flash steam. This is precisely what happens within thermodynamic traps as the hot condensate loses pressure in passing through the trap

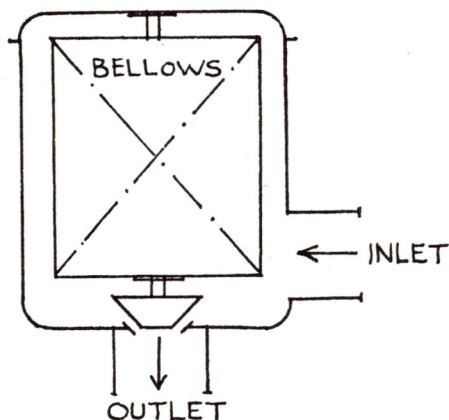

Fig. 7.4. Balanced pressure thermostatic trap.

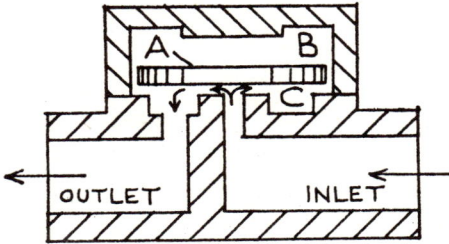

Fig. 7.5. Kinetic energy thermodynamic trap.

orifices. The degree of flashing determines the pressure set up in a control chamber located above the free valve, and hence whether the valve is open or closed. The comparatively recently developed kinetic energy thermodynamic trap, Fig. 7.5, is particularly suitable for use in inaccessible positions, since it has only one moving part and requires practically no maintenance. When cool condensate enters the trap the valve disc A is lifted away from the raised seats C, and condensate and air flow at low velocity to the outlet. As the system warms up, hotter condensate approaching steam temperature enters the trap and some re-evaporation takes place, with a resulting increase in flow velocity, a decrease in static pressure under the disc and an increase in pressure in chamber B. The pressure in B acts on the total area of the valve disc and eventually exceeds the combined pressure of the incoming hot condensate and the leaving condensate — flash steam mixture acting on the underside of the disc. The valve disc is therefore forced down and closes both the inlet and outlet ports. It will remain closed for as long as the pressure above the disc is higher than that below it. The chamber B has only a small volume, so the trapped flash steam quickly condenses and loses pressure, thus allowing the disc to rise and repeat the cycle. When handling condensate at constant temperature and pressure the time of the cycle will also be constant, and the flow will therefore take place in regular and rapid pulses.

The pressure at the outlet from any of the above traps, i.e. the pressure at the inlet minus the pressure drop across the trap, is available for lifting the condensate and overcoming the resistance offered to flow by the condensate piping. As it is difficult to determine the actual pressure loss due to the mixture of condensate, air and flash steam flowing in the condensate piping, it is usual to allow only 60 cm lift for every 10 kN/m^2 pressure at the trap outlet. If the pressure is too low to lift the condensate to the required height it is necessary to either collect the condensate at low level and then pump it up to the required height or, alternatively, use a lifting or pumping trap. Such traps are particularly suitable for handling hot condensate, which would otherwise give trouble due to flashing at the pump suction. These traps use the mechanical float principle and a high-pressure

TABLE 7.1
Main Features of Steam Traps

Type of trap	Discharge	Temperature of condensate	Discharges air	Withstand shock	Liable to freezing	Suitable for superheated steam	Suitable for varying pressure
Open-top bucket	Essentially intermittent	Saturation	Not very well, requires air vent	Yes	Yes	Yes	Yes
Inverted bucket	Intermittent, continuous during start-up	Saturation	Yes	Yes	Yes	Limited degree	Limited variation
Spherical free float	Continuous	Saturation	Yes	No	Some types	Yes	Yes
Ball valve, plain float	Continuous	Saturation	Thermostatic air release required	No	Yes	Yes	Yes
Slide-valve	Continuous	Saturation	Not very well, requires air vent	No	Yes	Yes	Yes
Liquid expansion	Semi-continuous (sluggish response to changing conditions)	Adjustable	Yes	Yes	No	Limited degree	No
Metallic expansion	Semi-continuous (sluggish response to changing conditions)	Adjustable	Yes	Yes	No	Limited degree	No
Tube expansion	Semi-continuous (sluggish response to changing conditions)	Adjustable	Yes	Yes	Yes	Limited degree	No
Balanced pressure (diaphragm or bellows)	Semi-continuous. Discharges condensate as it is formed after cooling	About 5°C below saturation	Yes	No	No	No	Yes
Thermodynamic	Rapid intermittent	Saturation	Yes	Yes	No	Yes*	Yes
Labyrinth	Semi-continuous impulse	Saturation	Yes	Yes	No	Yes	Yes
Lifting or pumping	Intermittent	Saturation and below	Not very well, requires air vent	Yes	Yes	Yes	Yes

* Minimum pressure about 70 kN/m^2

steam supply as follows. Low-pressure condensate collects in the fairly large body of the trap, and as the float rises it automatically opens the outlet valve and the high-pressure steam inlet valve. The low-pressure condensate is thus blown out of the trap, the lift being determined by the pressure of the high-pressure steam supply.

It is advisable to install a strainer at the inlet of all steam traps to reduce the risk of damage to the valve and seat by pipe scale.

When selecting steam traps from manufacturers' lists it is important to consider not only the steady condensation rate of the system but also the rate of condensation which takes place during warming up of the pipework and heating apparatus. The steam consumption during start-up from cold may be considerable, and the pressure at the trap inlet correspondingly low. These conditions may well determine the required duty of the trap.

The main features affecting the selection of steam traps are summarized in Table 7.1.

The properties of water in its liquid and vapour phases that have been used in the following examples are taken from the abridged set of steam tables included in *Thermodynamic and Transport Properties of Fluids,* Y. R. Mayhew and G. F. C. Rogers (Blackwell, 1964).

Example 7.1. Calculate the amount of saturated flash steam that would be produced at 0·14 MN/m² absolute pressure from 0·63 kg/s of condensate discharged at saturation temperature from process plant operating at 0·8 MN/m² absolute pressure.

From steam tables the properties of the condensate and flash steam are found to be:

$p,$ $MN/m^2 abs$	$h_f,$ kJ/kg	$h_{fg},$ kJ/kg	$h_g,$ kJ/kg
0·8	721	–	–
0·14	458	2 232	2 690

From which the surplus heat in the condensate is $721 - 458 = 263$ kJ/kg and the amount of flash steam formed will be:

$$\frac{263}{2\,232} \times 0{\cdot}63 = 0{\cdot}074 \text{ kg/s}$$

Alternatively, from Eq (7.1) and neglecting in this case the live-steam make-up term x:

$$y \cdot h_{f1} = z h_{g2} + (y - z) h_{f_2}$$

that is $\qquad 0{\cdot}63 \times 721 = 2\,690z + (0{\cdot}63 - z)458$

or $\qquad\qquad\qquad 2\,232z = 165{\cdot}6$

and $\qquad\qquad\qquad z = 0{\cdot}074$ kg/s as before.

The residual condensate at 0·14 MN/m² absolute pressure and saturation temperature will be

$$0·63 - 0·074 = 0·556 \text{ kg/s}$$

Example 7.2. A process heating plant uses the latent heat only from 0·25 kg/s of steam at 0·9 MN/m² absolute pressure and having a dryness fraction of 0·9. The condensate is discharged into a flash-steam vessel that supplies a low-pressure plant operating with dry saturated steam at 0·21 MN/m² absolute pressure. Determine how much live steam at 0·9 MN/m² absolute pressure should be passed through a steam-pressure reducing valve to make-up a total of 0·063 kg/s for the low-pressure plant.

From steam tables:

p MN/m²	h_f, kJ/kg	h_{fg}, kJ/kg	h_g, kJ/kg
0·9	743	2 031	—
0·21	511	2 198	2 709

The specific enthalpy of the wet steam having a dryness fraction of 0·9 will be:

$$743 + 0·9 \times 2\,031 = 2571 \text{ kJ/kg}$$

Then, using Eq (7.1),

$$2\,571x + 0·25 \times 743 = 0·063 \times 2\,709 + 511\,(x + 0·25 - 0·063)$$
$$1\,827·9x + 185·7 = 170·67 + 511x + 95·56$$
$$1316·9x = 80·53$$
$$x = 0·038 \text{ kg/s}$$

The residual condensate $= 0·038 + 0·25 - 0·063 = 0·225 \text{ kg/s}$

Thus, the amount of steam passing through the reducing valve is 0·061 kg/s. It should be noted that when steam passes through a reducing valve its specific enthalpy remains constant, but there is a change in the proportion of latent heat and sensible heat, depending upon the state of the high- and low-pressure steam. In this example the specific enthalpy of the steam after pressure reduction is 1 827.9 kJ/kg which is less than the specific enthalpy of dry saturated steam at the same pressure. Thus, the make-up steam is not dry saturated but in giving up its heat, together with that in the condensate from the high-pressure plant, the required amount of dry saturated steam at the lower pressure is produced. In practice, the live-steam make-up pipe is often connected into the flash-steam main after the flash-steam vessel. If this method had been adopted in this example dry saturated steam at the lower pressure would not have been produced.

Example 7.3. A certain process having a heat requirement of 220 kW uses the latent heat only of steam at 0·8 MN/m² absolute pressure and 0·9 dry. The condensate is led to a flash-steam vessel that supplies steam at 0·14 MN/m² absolute pressure and 0·9 dry to a low-pressure plant. In addition, live steam at the rate of 0·013 kg/s is passed through a reducing valve and enters the flash steam main after the flash vessel as shown in Fig. 7.6. Neglecting heat and pressure losses along the various pipes determine:

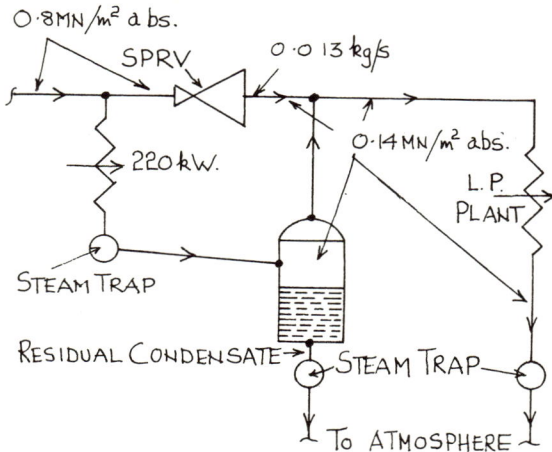

Fig. 7.6. Flash steam with live-steam make-up.

(*a*) the quality of the steam entering the low-pressure plant;

(*b*) the heat requirement of the low-pressure plant — assume that only the latent heat in the steam is used;

(*c*) the total amount of dry flash steam that would be formed if the residual condensate from the flash vessel and the condensate from the low-pressure plant are discharged to atmosphere.

From steam tables:

p, MN/m²	h_f, kJ/kg	h_{fg}, kJ/kg
0·8	721	2 048
0·14	458	2 232
atmos.	417	2 258

The specific enthalpy of the wet steam:

at 0.8 MN/m² abs pressure

$$= 721 + 0.9 \times 2\,048 = 2\,564$$

and at 0.14 MN/m² abs pressure

$$= 458 + 0.9 \times 2\,232 = 2\,467$$

(*a*) Amount of condensate at saturation temperature from the high-pressure plant

$$= \frac{220}{0.9 \times 2\,048} = 0.119 \text{ kg/s}$$

Then amount of flash steam formed at 0.14 MN/m² abs pressure and 0.9 dry

$$= \frac{0.119\,(721 - 458)}{0.9 \times 2\,232} = 0.016 \text{ kg/s}$$

Residual condensate from flash vessel

$$= 0.119 - 0.016 = 0.103 \text{ kg/s}$$

Total amount of steam at 0.14 MN/m² abs pressure supplied to low-pressure plant

$$= 0.013 + 0.016 = 0.029 \text{ kg/s}$$

having a specific enthalpy of

$$\frac{(0.013 \times 2\,564) + (0.016 \times 2\,467)}{0.029} = 2\,510 \text{ kJ/kg}$$

Since the specific enthalpy of the steam

$$= h_f + x \cdot h_{fg}$$

where x = dryness fraction

Then $2\,510 = 458 + 2\,232x$

from which $x = 0.92$

The steam entering the low-pressure plant is therefore 0.92 dry.

(*b*) The low-pressure plant condenses 0.029 kg/s of steam at 0.14 MN/m² absolute pressure and 0.92 dry. The heat requirement of this plant is therefore

$$0.029 \times 0.92 \times 2\,232 = 59.55 \text{ kW}$$

(*c*) Total amount of condensate at 0.14 MN/m² absolute pressure discharged to atmosphere will be:

$$0.103 + 0.029 = 0.132 \text{ kg/s}$$

The amount of flash steam formed at atmospheric pressure will be:

$$\frac{0{\cdot}132\,(458 - 417)}{2\,258} = 0{\cdot}0\,024 \text{ kg/s}$$

Example 7.4. The continuous blow-down from a steam boiler working at $1{\cdot}7$ MN/m² absolute pressure takes place at the rate of $0{\cdot}13$ kg/s and is passed to a flash steam vessel operating at $0{\cdot}21$ MN/m² absolute pressure and producing dry saturated steam. The residual blow-down from this vessel is then passed to a second flash-steam vessel that supplies steam at $0{\cdot}12$ MN/m² absolute pressure and $0{\cdot}9$ dry. Determine:

(*a*) How much of the original blow-down is available as boiler feed-water if only the condensed flash steam is collected for re-use.

(*b*) The total heat recovered from the flash steam produced by the two vessels. Assume that the heaters supplied use only the latent heat in the flash steam

From steam tables:

$p,$ MN/m^2	$h_f,$ kJ/kg	$h_{fg},$ kJ/kg
1·7	872	1 923
0·21	511	2 198
0·12	439	2 244

(*a*) Assuming that the continuous blow-down from the boilers is at saturation temperature, the amount of flash steam produced at $0{\cdot}21$ MN/m² absolute will be:

$$\frac{0{\cdot}13\,(872 - 511)}{2\,198} = 0{\cdot}021 \text{ kg/s}$$

The residual blow-down passed to the second vessel will be:

$$0{\cdot}13 - 0{\cdot}021 = 0{\cdot}109 \text{ kg/s}$$

and the amount of flash steam produced at $0{\cdot}12$ MN/m² absolute pressure and $0{\cdot}9$ dry will be:

$$\frac{0{\cdot}109\,(511 - 439)}{0{\cdot}9 \times 2\,244} = 0{\cdot}004 \text{ kg/s}$$

The amount of original blow-down available as boiler feed water will be:

$$0{\cdot}021 + 0{\cdot}004 = 0{\cdot}025 \text{ kg/s}$$

and the amount passed to drain will be

$$0{\cdot}13 - 0{\cdot}025 = 0{\cdot}105 \text{ kg/s}$$

(*b*) Amount of heat recovered from the flash steam at 0.21 MN/m² absolute pressure

$$= 0.021 \times 2\,198 = 462 \text{ kW}$$

and from the flash steam at 0.12 MN/m² absolute pressure

$$= 0.004 \times 0.9 \times 2\,244 = 8.1 \text{ kW}$$

Making a total heat recovery of

$$46.2 + 8.1 = 54.3 \text{ kW}$$

Example 7.5. A steam boiler operating at 0.7 MN/m² absolute pressure supplies dry saturated steam to a process plant that uses the latent heat in the steam only and discharges the condensate at saturation temperature to a flash vessel that supplies wet steam at 0.12 MN/m² absolute pressure. Make-up steam at the rate of 0.016 kg/s passes through a reducing valve and also enters the flash vessel, which supplies a total of 0.065 kg/s of dry saturated steam at 0.12 MN/m² absolute pressure. Subsequently the boiler is required to work at a pressure of 0.35 MN/m² absolute. If the heat requirements remain the same:

(*a*) Determine the amount of condensate in the first instance and under changed conditions.

(*b*) How much live-steam make-up will be necessary under changed conditions.

From steam tables:

p, MN/m^2	h_f, kJ/kg	h_{fg}, kJ/kg	h_g, kJ/kg
0.7	697	2 067	2 764
0.35	584	2 148	2 732
0.12	439	2 244	2 683

(*a*) Then, using Eq (7.1),

$$0.016 \times 2\,764 + 697y = 0.065 \times 2\,683 + 439\,(0.016 + y - 0.065)$$

from which the amount of condensate in the first instance will be:

$$y = 0.421 \text{ kg/s}$$

Therefore the heat supplied to the higher pressure plant

$$= 0.421 \times 2\,067 = 870.2 \text{ kW}$$

Since this remains constant, then the amount of condensate under changed conditions will be

$$\frac{870.2}{2\,148} = 0.405 \text{ kg/s}$$

(*b*) Using Eq (7.1),

$$2\,732x + 0{\cdot}405 \times 584 = 0{\cdot}065 \times 2\,683 + 439\,(x + 0{\cdot}405 - 0{\cdot}065)$$

from which

$$x = 0{\cdot}038 \text{ kg/s}$$

Example 7.6. An industrial steam heating installation serves three separate processes having the following requirements:

Process A: 880 kW using steam at $1{\cdot}0$ MN/m^2 absolute pressure and 0·8 dry.

Process B: 590 kW using steam at $0{\cdot}7$ MN/m^2 absolute pressure and 0·9 dry.

Process C: 300 kW using steam at $0{\cdot}5$ MN/m^2 absolute pressure and dry saturated.

The condensate from these processes is discharged into a common flash vessel that supplies dry saturated steam at $0{\cdot}14$ MN/m^2 absolute pressure. The condensate from each process and the residual condensate from the flash vessel cools 3°C due to heat losses from the pipes. The residual condensate is discharged into an open hot-well.

(*a*) How much flash steam is formed?

(*b*) If the flash steam is used as the primary heating medium in a steam to water heat exchanger from which the condensate is discharged at 80°C, how much heat is recovered from the flash steam?

(*c*) How much residual condensate is discharged to the hot-well and how much will form flash steam at atmospheric pressure?

From steam tables:

p, MN/m^2	h_f, kJ/kg	h_{fg}, kJ/kg
1·0	763	2 015
0·7	697	2 067
0·5	640	2 109
0·14	458	2 232
atmos	417	2 258

Assume that the specific heat capacity of condensate = 4·2 kJ/kg °C

(*a*) Amount of condensate from:

$$\text{Process A} = \frac{880}{0{\cdot}8 \times 2\,015} = 0{\cdot}546 \text{ kg/s}$$

$$\text{Process B} = \frac{590}{0{\cdot}9 \times 2\,067} = 0{\cdot}317 \text{ ,,}$$

$$\text{Process C} = \frac{300}{2\,109} = 0{\cdot}142 \text{ ,,}$$

$$\overline{1{\cdot}005 \text{ ,,}}$$

Amount of flash steam formed at $0.14\ \text{MN/m}^2$ absolute pressure from the condensate from:

$$\text{Process A} = \frac{0.546\,(763 - 3 \times 4.2 - 458)}{2\,232} = 0.072\ \text{kg/s}$$

$$\text{Process B} = \frac{0.317\,(697 - 3 \times 4.2 - 458)}{2\,232} = 0.032\ \text{''}$$

$$\text{Process C} = \frac{0.142\,(640 - 3 \times 4.2 - 458)}{2\,232} = 0.011\ \text{''}$$

$$\overline{0.115\ \text{''}}$$

Residual condensate from flash vessel

$$= 1.005 - 0.115 = 0.89\ \text{kg/s}$$

(*b*) Heat recovered from $0.115\ \text{kg/s}$ of dry saturated steam at $0.14\ \text{MN/m}^2$ absolute pressure when discharged at $80°\text{C}$

$$= 0.115\,(2\,232 + 458 - 80 \times 4.2)$$
$$= 270.7\ \text{kW}$$

(*c*) Amount of residual condensate passed to hot-well = $0.89\ \text{kg/s}$ and amount of flash steam formed at atmospheric pressure

$$= \frac{0.89\,(458 - 417)}{2\,258} = 0.016\ \text{kg/s}$$

Example 7.7. A heater battery condenses $0.065\ \text{kg/s}$ of steam at $0.93\ \text{MN/m}^2$ absolute pressure and discharges the condensate at saturation temperature to a flash vessel located 10 m above the level of the steam trap. What is the maximum pressure at which flash steam will be formed and how much flash steam will be formed at this pressure.

To allow for the pressure drop across the steam trap and along the rising pipe for two-phase flow conditions, it is usual to allow $10\ \text{kN/m}^2$ for every 60 cm of lift. The maximum pressure at the flash vessel will therefore be:

$$0.93 - \frac{10 \times 0.01}{0.6} = 0.764\ \text{MN/m}^2$$

For steam tables:

p, MN/m^2	h_f, kJ/kg	h_{fg}, kJ/kg
0.93	749	–
0.764	712	2 060

The amount of flash steam formed will be:

$$\frac{0.065\,(749 - 712)}{2\,060} = 0.0012\ \text{kg/s}$$

Example 7.8. List the main requirements of a flash-steam system and give a sketch showing a typical flash-steam vessel and its fittings.

1. The pressure of the flash steam should be as low as possible for the following reasons:

(i) to ensure maximum heat recovery;
(ii) to minimize further flashing of the residual condensate at the open hot-well;
(iii) to increase thermal efficiency, particularly when used in conjunction with exhaust steam.

2. The amount of flash steam formed should preferably be less than the demand with the deficit made up from live steam via a pressure reducing valve. If the flash is greater than the demand the pressure in the flash vessel will rise and steam will be lost via the safety valve.

3. The dimensions of the flash vessel should be such that the steam velocity is low and not greater than 10 m/s in the outlet pipe. This allows separation of the steam from the condensate and keeps carry-over of water droplets to a minimum, thus producing fairly dry steam.

4. Wherever practicable the flash steam should be used locally. In this way heat losses from the attendant pipe work are kept to a minimum, thus ensuring maximum heat recovery.

5. The recovery and use of flash steam require extra plant and equipment, and therefore extra expenditure. The demand for low-pressure steam should therefore coincide with the discharge of high-pressure condensate. If this is not practicable it may be more economical to use all live steam via a pressure-reducing valve.

6. The flash-steam vessel should be fitted with a pressure gauge and safety valve and either a steam trap or float-operated valve to ensure rapid removal of the residual condensate. A typical flash-steam vessel in shown in Fig. 7.7.

Example 7.9. The discharge from a process heating plant working at 0·6 MN/m² abs. pressure and 20 per cent moisture content provides sufficient flash steam at 0·11 MN/m² abs. pressure and 10 per cent moisture content to operate a heating system emitting 20 kW. If residual condensate leaves the flash vessel at the rate of 0·065 kg/s, determine how much live steam passes through the steam trap.

From steam tables:

p, MN/m^2	h_f, kJ/kg	h_{fg}, kJ/kg
0·6	670	2 087
0·11	429	2 251

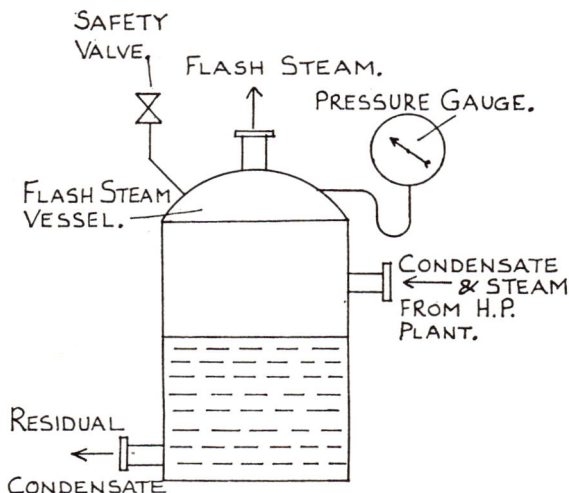

Fig. 7.7. A typical flash-steam vessel.

Amount of flash steam formed

$$= \frac{20}{0.9 \times 2\,251} = 0.01 \text{ kg/s}$$

Use Eq (7.1), but in this case let x = amount of steam passing through the steam trap.

$(670 + 0.8 \times 2\,087)x + 670y = 0.01(429 + 0.9 \times 2\,251) + 0.065 \times 429$

that is $3.49x + y = 0.078$

also $x + y = 0.065 + 0.01 = 0.075$

By subtraction $2.49x = 0.003$

and $x = 0.0012 \text{ kg/s}$

Example 7.10. Briefly describe and illustrate the following steam heating systems: (*a*) one-pipe; (*b*) two-pipe; (*c*) exhaust.

(*a*) With one-pipe systems steam and condensate flow in the same pipe, sometimes in the same direction and sometimes in opposite directions. As a result of this the system is used only for small, usually radiator, heating systems. Fig. 7.8 (*a*) shows the simplest arrangement in which the condensate from the first riser drains back against the steam flow into the steam main and then flows forward with the steam for the second riser. For larger installations it is advisable to drain each riser or drop as shown at (*b*) and (*c*). With arrangements (*a*) and (*b*) the steam main should be

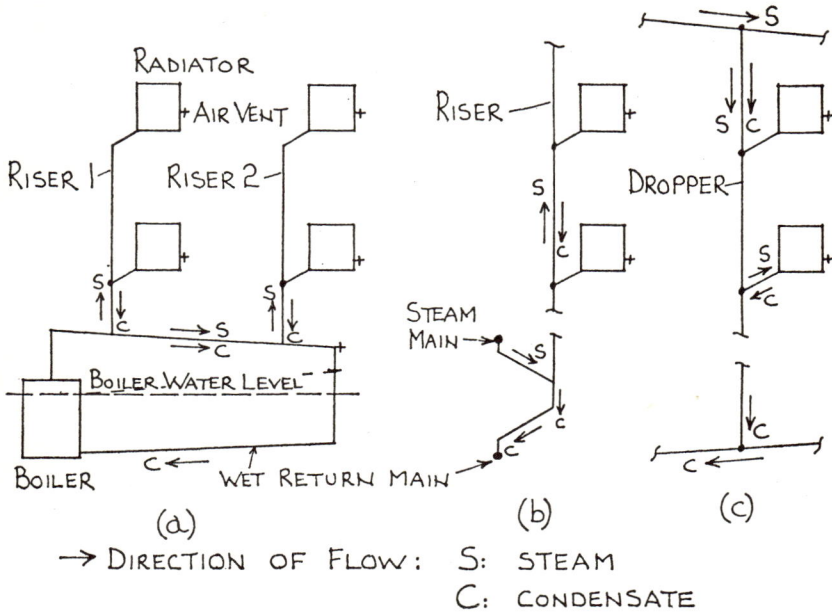

Fig. 7.8. Steam systems, basic arrangement.

drained and air vented after feeding the last riser. Depending on whether the building has a basement or not, the condensate return main may be run below the level of the water in the boiler as a "wet" return or above the water level in the boiler as a "dry" return. Wherever practicable a wet return arrangement should be adopted in preference to dry returns which, unless each riser is provided with a water seal, may become steam locked.

These small one-pipe systems usually operate in totally enclosed conditions, with the condensate returning to the boiler by natural circulation. With steam flowing and condensing there is a pressure difference across the system which is balanced by an equivalent rise in the water level in the return drops. Any tendency to raise this level above that required for equilibrium results in a positive water flow towards the boiler. In this way the condensate returning from the radiators circulates back to the boiler for re-use. When starting up from cold the pressure drop is greater than during normal operation, and the rise in water level in the return drop is correspondingly greater. To prevent water from leaving the boiler at these times a non-return valve may be fitted into the main near the boiler. These systems operate at pressures up to about 0.12 MN/m^2 absolute.

Each radiator is fitted with a thermostatic air vent of either the pressure or vacuum type. The latter enabling the system to operate under vacuum conditions by preventing air from re-entering the system during periods of cooling down.

(*b*) With two-pipe systems the steam and condensate flow in separate pipes. Such systems are suitable for both small and large installations and may be arranged to operate at sub-atmospheric, low or high pressure.

Fig. 7.9 illustrates a typical two-pipe system arranged for combined space and process heating at low and high pressures respectively.

The efficiency of the cycle may be improved by installing a flash condenser and economiser in the condensate return mains. Alternatively, flash steam from the higher-pressure condensate may be used for the lower-pressure equipment.

If sufficient pressure is available at the outlet of the steam traps the condensate may be returned in mains located at high level. With some large low-pressure installations it may be more convenient to collect all the condensate in a receiver at low level and then to return it to the boiler house by a float-operated pump.

With all two-pipe systems no connection should be made between the steam and condensate mains except through a steam trap.

For fairly large installations operating at pressures only slightly above atmospheric the flow of steam may be assisted by using a closed system and a vacuum pump in the return main for removing condensate from the system. With these systems thermostatic type steam traps and packless gland type valves would normally be used. The vacuum return main system provides positive steam flow and condensate removal and is suitable for installations in which temperature modulation is not a primary requirement. For space-heating installations that require full control of the building

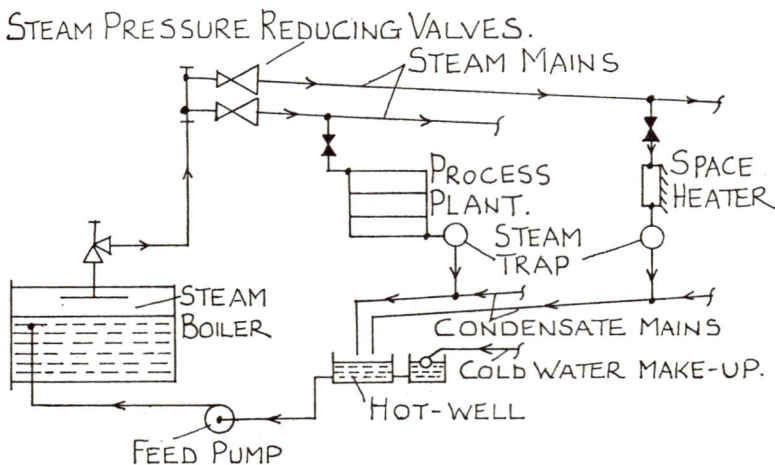

Fig. 7.9. Two-pipe combined space and process heating.

temperature at varying external temperatures it is necessary to vary automatically the temperature of the steam, and hence heat emission from the
heating appliances. This may be done by modifying the vacuum return
main system to include automatic control of the steam pressure so that
sub-atmospheric pressures may be maintained in both steam and condensate mains. The steam-supply control valve may be operated by room
thermostat alone or by room thermostat combined with an outdoor
temperature sensing element. These systems operate at pressures above
atmospheric only during very cold spells, during mild spells the steam
pressure may be maintained as low as $0.05 \, MN/m^2$ abs. Steam flow and
condensate removal are maintained at all times by controlling the vacuum
pump to maintain a fixed pressure differential across the system. The main
features of a differential vacuum steam system are shown in Fig. 7.10.

(c) Exhaust steam from non-condensing engines may be used for heating purposes either by passing it through a steam-to-water heat exchanger
and circulating the heated water through the heating system or by using
it directly. In both cases care must be taken to ensure that the permissible
back pressure on the engine is not exceeded. This may be done by
installing an exhaust relief valve in the main exhaust header. Exhaust
steam is invariably contaminated with engine lubricating oil and possibly

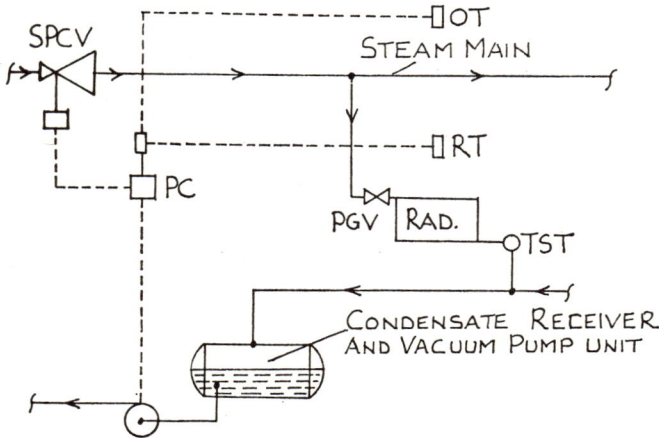

OT : Outside Thermostat
RT : Room Thermostat
PC : Pressure Controller
SPVC: Steam Pressure Control Valve
PGV : Packless Gland Valve
TST : Thermostatic Steam Trap

Fig. 7.10. Differential vacuum steam system.

with small droplets of water of condensation, and must therefore be passed through an oil separator before it is used in any form of heating apparatus. A diagram showing the main features of an exhaust steam system is given in Fig. 7.11.

At times there may be insufficient exhaust steam to meet the secondary heating requirements, and when this occurs live steam direct from the boilers may be fed to the heating system through a steam-pressure reducing valve. At times when the exhaust steam is not required it may be passed to a heat accumulator or condensed in the usual way. The use of exhaust steam is conducive to fuel economy, and should always be considered. Since the exhaust steam is at low pressure, it may often be used in conjunction with a vacuum return main system.

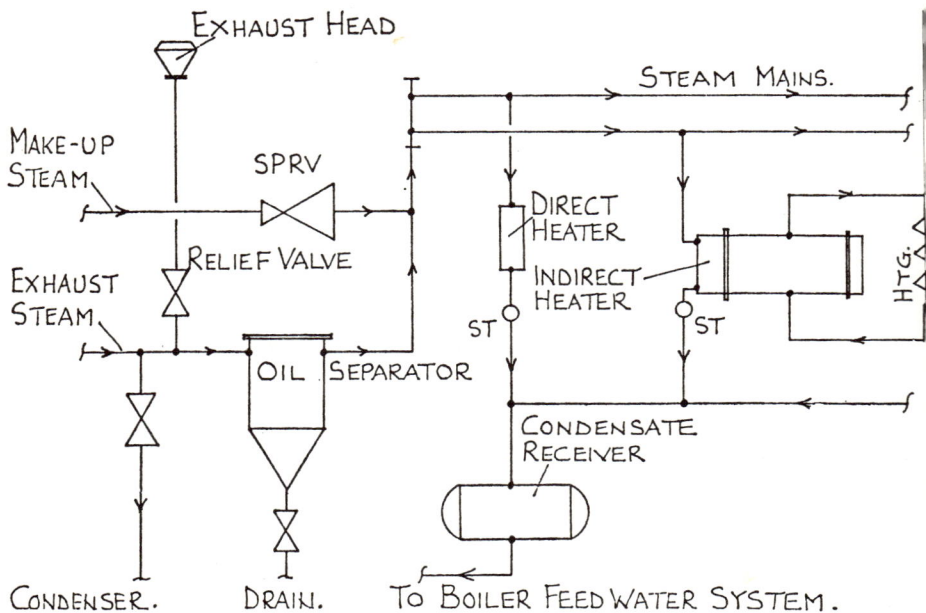

Fig. 7.11 Exhaust steam system.

Problems

1. Calculate the amount of dry saturated flash steam produced at 0.14 MN/m^2 absolute pressure from 9 000 kg of condensate at 0.8 MN/m^2 absolute pressure.

Ans.: 1 060 kg.

2. A heater battery having an output of 220 kW is supplied with steam at 0·4 MN/m² abs. pressure and 80 per cent dry. The condensate from the battery is passed to a flash steam vessel which supplies steam at 0·14 MN/m² abs. pressure and 90 per cent dry. Determine how much flash steam is formed. (IHVE)

Ans.: 0·0094 kg/s;

3. Flash steam at a rate of 0·004 kg/s is discharged from a flash-steam vessel working at 0·48 MN/m² abs. pressure and is used in a hot-water service calorifier as the primary heating medium. The condensate leaves the calorifier at 93·3°C. Assume that the specific heat capacity of the condensate is 4·2 kJ/kg °C.

(*a*) What is the amount of heat recovered from the flash steam?

(*b*) The flash steam is recovered from condensate discharged at 1·1 MN/m² abs. pressure. If the steam trap through which it passes leaks and the weight of live steam that escapes is 5 per cent of the weight of the genuine condensate, what is the amount of genuine condensate?

(*c*) After repair to the trap, how much flash steam will be available?

Ans.: (*a*) 9·43 kW; (*b*) 0·033 kg/s; (*c*) 0.0023 kg/s. (IHVE)

4. Steam heaters supplied from three steam mains at different pressures and moisture contents as shown below drain, through traps of the same construction and resistance to flow, into a common high-level condensate main which delivers the condensate into a flash vessel. The lift is 5 m.

(*a*) What is the highest possible pressure in the flash steam vessel?

(*b*) What is the amount of flash steam formed?

(*c*) Make a sketch of the arrangement.

Data:

Group 1: heater output 87·9 kW; 0·59 MN/m² abs. and 20 per cent moisture.

Group 2: heater output 146·6 kW; 0·48 MN/m² abs. and 10 per cent moisture.

Group 3: heater output 175·9 kW; 0·38 MN/m² abs. and dry saturated. (NC)

Ans.: (*a*) 0·214 MN/m² abs. (*b*) 0·011 kg/s

5. A steam plant operates at 0·93 MN/m² abs. pressure and is equipped with a flash-steam recovery vessel of 0·3 m internal diameter, capable of generating 2·44 kg/gm² wet steam at 0·14 MN/m² abs. pressure and 7 per cent moisture content. On the low-pressure side of the steam installation

0·315 kg/s dry saturated steam is required. How much high-pressure condensate is needed to produce the rated amount of flash steam and how much live steam has to be supplied to make up the required low-pressure steam.

Sketch the plant arrangement showing pipe lay-out and recovery vessel accessories.   (NC)

Ans.: 1·23 kg/s and 0·15 kg/s

6. A factory uses the latent heat only from 0·13 kg/s of dry saturated steam at 0·8 MN/m² abs. pressure for process heating. The condensate is discharged into a flash-steam vessel that supplies a low-pressure steam line operating at 0·21 MN/m² abs. pressure. How much live dry saturated steam at 0·8 MN/m² abs. pressure is required to make up a total of 0·038 kg/s dry saturated steam in the low-pressure steam line? Neglect heat losses from the vessel and attendant pipe work.   (IHVE)

Ans.: 0·025 kg/s.

7. Plot a curve showing the percentage weight of water converted to saturated flash steam when condensate at steam temperatures and pressures up to 1·48 MN/m² absolute is discharged to a flash-steam recovery system working at 0·17 MN/m² absolute pressure.   (IHVE)

8. Select a steam trap for handling the condensate from: (*a*) a warm-air heater battery, and (*b*) a steam-heated radiator. Give reasons for your choice.

Ans.: (*a*) Select a steam trap that will discharge the condensate as soon as it is formed, thus preventing the battery from becoming flooded.

(*b*) The trap should be small and suitable for handling varying rates of condensation and pressure changes.

9. The condensate returning to the hot well in a boiler house derives from a high-pressure steam installation operating at 0·96 MN/m² abs. pressure. The delivery steam carries 22°C of superheat, and the total steam requirements, including losses, amount to 900 kW. The condensate returning to the hot well passes through a cascade heater to heat low-pressure hot water. 98 kJ/kg has been lost between the traps and the cascade heater.

(*a*) Find the amount of hot water which can be generated in the cascade heater for the low-pressure hot-water installation, the flow and return temperatures of which are 82·2°C and 71·1°C respectively.

(*b*) State the net hourly amount of heat required to generate steam from the condensate returning from the cascade heater (assume no losses).

(*c*) Sketch the arrangement of the suggested cascade hot-water-heating plant.   (IHVE)

(Take the specific heat capacity of the steam and water as 2·1 and 4·2 kJ/kg °C respectively)

Ans.: (*a*) 2·9 kg/s

(*b*) 1 077 kW.

(*c*) Arrange for condensate to mix with returning water from heating system and also automatic control of water level in cascade heater vessel.

8: Steam Pipe Sizing

Introduction

The pressure at which steam should be conveyed from the boilers to the point of use depends on the circumstances of each individual case. There are, however, two main alternatives:

(a) Generation and distribution at high pressure with steam-pressure reducing valves located in the branches to the heating units. This arrangement would be particularly suitable where a number of process-heating appliances operating at different pressures and temperatures are required to be supplied with steam from a common boiler plant. Also, operating the boiler at pressures in excess of the pressure required by the steam users provides some steam storage, thus enabling the system to operate under fluctuating load conditions and to cope with short periods of heavy demand. While the use of high pressures results in smaller pipe diameter, high-pressure plant and equipment is generally more expensive, and maintenance costs may be higher than for low-pressure plant. The generation and distribution pressure, therefore, depends upon economic considerations that should be worked out before a decision is made.

(b) Generation and distribution at the temperature and pressure required by the heating units, with due allowance being made for pressure drop along the mains. This arrangement would be used for most space-heating systems operating at comparatively low pressure.

The flow of steam is automatically induced by the volume shrinkage that occurs as the steam gives up its latent heat and condenses. There is no need therefore to provide a "head", as with water systems, to maintain flow.

When determining the size of a steam main the following information is normally required:

(a) initial or final steam pressure, temperature and quality;
(b) steam flow rate based on the heat requirement of the heating units and any waste heat from the distribution piping;
(c) length of pipe;
(d) permissible pressure drop;
(e) permissible velocity of flow.

Items (*d*) and (*e*) are interdependent and depend on several factors such as:

(*a*) Relative direction of steam and condensate flow where they occur within the same pipe.

(*b*) Whether the pipe is vertical, horizontal or sloping down in the direction of steam flow or against it.

(*c*) Steam quality and the erosive action of wet steam on valve seats.

(*d*) The possibility of carry-over of water droplets from boiler steam spaces and flash steam vessels into the distribution piping.

(*e*) Permissible noise level.

The pressure drops and velocities given in Tables 8.1 and 8.2 may be used as a guide.

TABLE 8.1

Typical Pressure Drops

Small one- and two-pipe closed systems, including vacuum systems, for space heating: up to $20\frac{N}{m^2}\Big/m$

Open systems:

(*a*) Low pressure: up to 200 kN/m² abs; up to $200\frac{N}{m^2}\Big/m$

(*b*) Medium pressure: 200 to 300 kN/m² abs; $200\frac{N}{m^2}\Big/m$

(*c*) High pressure: above 300 kN/m² abs; $200\frac{N}{m^2}\Big/m$.

TABLE 8.2

Typical Steam Velocities

Saturated steam: up to 40 m/s
Superheated steam: 40 – 65 m/s
Exhaust steam (wet): up to 30 m/s
Flash steam, exit from vessel: up to 15 m/s
Boiler outlets: 6 – 20 m/s
Steam headers: 10 – 25 m/s
Horizontal mains: Small closed systems with –

(*a*) steam and condensate flowing in opposite directions: 3 – 10 m/s
(*b*) steam and condensate flowing in same direction: 10 – 25 m/s

Risers:

(*a*) one-pipe systems: 5 – 10 m/s
(*b*) two-pipe systems: 6 – 15 m/s

For a steam system

$$\dot{m}(h_1 - h_2) = \Phi_p + \Phi_u \, ,$$

and $$\dot{m} = (\Phi_p + \Phi_u)/(h_1 - h_2) \qquad (8.1)$$

where

$\dot{m}$ = steam flow rate
Φ_p = heat emission from the piping
Φ_u = heat emission from the heating units
h_1 and h_2 = specific enthalpy at entry to the main and exit from the unit respectively.

For most insulated steam mains it is usual to assume that

$$\Phi_p = 0.10\Phi_u \text{ to } 0.15\Phi_u$$

and that $$h_1 - h_2 = h_{fg}$$

Eq (8.1) then becomes

$$\dot{m} = \frac{f\Phi_u}{h_{fg}} \qquad (8.2)$$

where $$f = 1.10 - 1.15$$

and h_{fg} = latent heat of steam.

If it is decided to use a definite velocity the diameter of the pipe may be determined as follows:

Let $\dot{m}$ = rate of steam flow, kg/s
v_g = specific volume of steam, m^3/kg
u = velocity, m/s
d = internal diameter of pipe, m

Then $$\frac{\pi d^2 u}{4} = \dot{m} v_g$$

from which $$d = 1.128 \sqrt{\frac{\dot{m} v_g}{u}} \qquad (8.3)$$

and $$u = \dot{m} v_g A^{-1} \qquad (8.4)$$

where A = cross section area of pipe.

Values of A^{-1}, based on the nominal diameter of heavy grade tubes to B.S. 1387 of 1967, are given in Table 8.3.

TABLE 8.3

Velocity Factors

Nominal diameter, mm	A m^2	A^{-1} m^{-2}
15	0.0001767	5 700
20	0.0003142	3 200
25	0.0004909	2 000
32	0.0008042	1 200
40	0.001257	800
50	0.001963	500
65	0.003318	300
80	0.005027	200

Note that velocity depends on the specific volume, and therefore on the pressure of the steam.

The approximate pressure drop due to friction in steel tubes may be obtained from tables given in *Guide to Current Practice, 1970*. These tables are based on the following formula, and are suitable for steam flowing at pressures between 0.1 MN/m² and 1.0 MN/m² abs, with velocities ranging from 5 to 50 m/s.

$$Z_1 - Z_2 = \frac{\dot{m}^{1.889} l}{2.378 \times 10^6 \, d^{5.027}} \qquad (8.5)$$

where Z_1 = initial pressure factor, $p_1^{1.929}$
Z_2 = final pressure factor, $p_2^{1.929}$
p = absolute pressure, bars.
$\dot{m}$ = steam flow rate, kg/s
d = internal diameter of pipe, m
l = length of pipe run, m

Values of Z for a few selected pressures are given in Table 8.4.

The loss of pressure due to local resistances such as bends, tees and valves is related to the velocity pressure and may be expressed as the length of straight pipe that would have the same pressure loss. Equivalent lengths (l_e values) for the loss of one velocity pressure are given in Table 8.5 and typical velocity pressure factors (K values) in Table 4.4.

Condensate return mains usually handle a mixture of air, steam and condensate in widely varying amounts. Since also the flow may be either continuous or intermittent, it is not possible to calculate the pressure drop or mean velocity and the following diameters are recommended:

Diameter of steam main, mm	20	25	32	40	50	65	80	100	125	150
Diameter of condensate main, mm	15	20	20	25	32	32	40	65	65	80

Tables 8.4, 8.5 and 4.4 contain data from *Guide to Current Practice, 1970* and will be used in the examples that follow.

TABLE 8.4

Pressure Factors for Compressible Flow

p = pressure MN/m^2 abs.

$Z = p^{1.929}$

p	Z	p	Z
0.12	1.42	0.43	16.67
0.13	1.66	0.44	17.43
0.15	2.19	0.45	18.20
0.17	2.78	0.46	18.99
0.29	7.80	0.47	19.79
0.30	8.32	0.48	20.61
0.31	8.87	0.49	21.45
0.40	14.50	0.50	22.30
0.41	15.21	0.51	23.17
0.42	15.93	0.52	24.05

In cases there thermostatic steam traps and air valves are used at the heaters, only water flows in the condensate main, and pipe diameters may be determined as for hot water.

Example 8.1. Fig. 8.1 shows the general layout of a steam distribution system for a factory centre erected on a sloping site. Each building requires 880 kW for space heating at a maximum flow-water temperature of 80°C and 0.32 l/s at 65°C for hot-water services supplied from separate steam to water heat exchangers. Give reasons for your recommendations regarding the steam pressure to be maintained in the boiler, steam mains and heat exchangers, and state what arrangements could be made for returning the

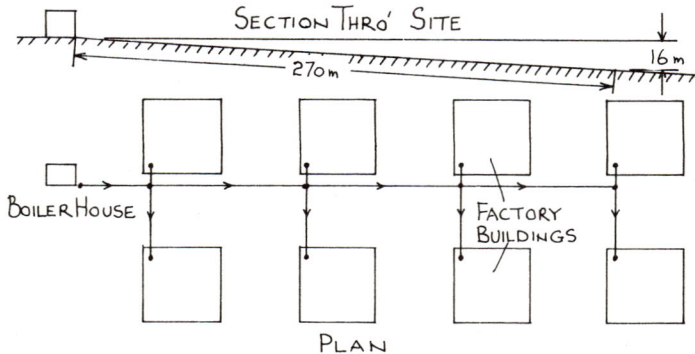

Fig. 8.1. General layout of a steam distribution system.

TABLE 8.5

Flow of Saturated Steam in Heavy Grade Black Steel Tubes to B.S. 1387: 1967

Z_1-Z_2	25		32		40		50		65		80	
l	$\dot{m}$	l_e	$\dot{m}$	l_e	$\dot{m}$	l_e	$\dot{m}$	l_e	$\dot{m}$	l_e	$\dot{m}$	l_e
0.0040	0.0075	0.8	0.0162	1.2	0.0247	1.5	0.0470	2.1	0.0957	3.0	0.1490	3.7
0.0063	0.0095	0.9	0.0206	1.3	0.0314	1.6	0.0598	2.2	0.1218	3.1	0.1895	3.8
0.0080	0.0108	0.9	0.0234	1.3	0.0357	1.6	0.0679	2.2	0.1382	3.1	0.2150	3.9
0.0100	0.0121	0.9	0.0263	1.3	0.0402	1.6	0.0764	2.2	0.1555	3.2	0.2420	3.9
0.0130	0.0139	0.9	0.0303	1.3	0.0461	1.6	0.0877	2.3	0.1787	3.2	0.2780	4.0
0.0160	0.0155	0.9	0.0338	1.3	0.0515	1.7	0.0979	2.3	0.1994	3.3	0.3103	4.1
0.0200	0.0175	0.9	0.0380	1.4	0.0579	1.7	0.1102	2.3	0.2245	3.3	0.3493	4.1
0.0250	0.0197	0.9	0.0428	1.4	0.0652	1.7	0.1240	2.3	0.2526	3.3	0.3930	4.2
0.0320	0.0224	1.0	0.0487	1.4	0.0743	1.7	0.1413	2.4	0.2879	3.4	0.4479	4.2
0.0400	0.0252	1.0	0.0549	1.4	0.0836	1.8	0.1591	2.4	0.3240	3.4	0.5041	4.3
0.0500	0.0284	1.0	0.0617	1.4	0.0941	1.8	0.1790	2.4	0.3646	3.5	0.5673	4.3
0.0630	0.0321	1.0	0.0698	1.5	0.1064	1.8	0.2023	2.5	0.4120	3.5	0.6411	4.4
0.0800	0.0364	1.0	0.0792	1.5	0.1207	1.8	0.2296	2.5	0.4676	3.6	0.7275	4.5
0.1000	0.0410	1.0	0.0891	1.5	0.1359	1.8	0.2584	2.5	0.5262	3.6	0.8188	4.5

Note. $\dot{m}$ = steam flow rate kg/s.
  l_e = equivalent length of pipe in metres when K = 1.0.
  Z = pressure factor, see Table 8.4.
  l = total length of pipe run in metres.

condensate to the boiler house for use as boiler feed water. Assume that the minimum water pressure in the hot-water heating and supply systems is 0.2 MN/m² abs.

The following main points should be considered:

1. The possibility of the secondary water boiling, on say breakdown of controls or failure of the circulating pumps, should be avoided by fixing the maximum steam temperature at, say, 5°C below the boiling point corresponding to the minimum pressure in the water systems. From steam tables the boiling point of water at 0.2 MN/m² abs is found to be 120°C. The maximum steam temperature would then be 120 − 5 = 115°C, corresponding to a steam pressure of 0.17 MN/m² abs. Each factory building should be supplied with steam at this temperature and pressure through a steam-pressure reducing valve in each branch.

Since the heating surface area, physical dimensions and cost of the heat exchangers depend upon the logarithmic mean temperature difference between the steam and the water, the steam temperature should be at least 10 to 20°C above the maximum water temperature. In this example the minimum difference is 115 − 80 = 35°C, which should result in an economically sized heat exchanger.

2. By keeping the pressure in the main fairly high, say 0.3 MN/m² abs, the pipe sizes required will be comparatively small. A typical pressure drop for long steam mains is $200 \dfrac{N}{m^2}\Big/m$ of equivalent length. Allowing, say, 30 per cent extra length for local resistances, the pressure drop along the main will be

$$270 \times 1.3 \times 200 \times 10^{-6} = 0.07 \ MN/m^2$$

The boiler would be required to operate at 0.37 MN/m² abs pressure.

3. When sizing the pipes, steam velocities in excess of 40 m/s should be avoided.

4. A dryness fraction of 0.9 should be assumed when calculating the steam flow rates. For steam at 0.17 MN/m² abs pressure the latent heat is found from steam tables to be 2 216 kJ/kg. Then, assuming a cold-water temperature of 10°C for the hot-water supply system and taking the specific heat capacity and density of the water as 4.2 kJ/kg°C and 1 000 kg/m³ respectively, the steam required by each factory building will be

$$\frac{880 + 0.32 \times 4.2(65 - 10)}{0.9 \times 2\,216} = 0.48 \ kg/s$$

Since heat losses from the various steam and water piping systems are unknown, it would be advisable to allow an extra 10 per cent on the steam demand when selecting a boiler. The minimum boiler output would be

$$1.1 \times 8 \times 0.48 = 4.224 \ kg/s \ at \ 0.37 \ MN/m^2 \ abs \ pressure$$

For the purpose of selecting a boiler from manufacturers' lists this output should be expressed as an equivalent evaporation of water from and at 100°C.

If $\dot{m}$ = actual evaporation, kg/s

h_g = specific enthalpy of the saturated vapour at the working pressure, kJ/kg

t_f = temperature of boiler feed water, °C.

Then

$$\text{Equivalent evaporation} = \frac{\dot{m}(h_g - 4.2t_f)}{2\,258} \ kg/s$$

where 2 258 kJ/kg is the latent heat of steam at 100°C.

Then, using the appropriate data from the steam tables and assuming a boiler feed water temperature of 80°C

$$\text{Equivalent evaporation} = \frac{4.224(2\,735 - 4.2 \times 80)}{2\,258}$$

$$= 4.5 \ kg/s \ from \ and \ at \ 100°C$$

5. The steam pressure at the inlet to the steam traps draining the heat exchangers is insufficient to lift and return the condensate back to the boiler house. This is particularly true at the lowest end of the site. [See Chapter 7, Example (7.7).] It will be necessary therefore to either drain the condensate to a low-level receiver and then pump it back to the boiler house or use lifting traps supplied with high-pressure steam direct from the main. In the second case it will be necessary to increase the quantity of steam generated by an amount equal to the steam demand of the lifting traps.

Example 8.2. A heavy grade steam main conveys 0.063 kg/s of saturated steam at an initial pressure of 0.17 MN/m² abs. The main is 50 m and contains 6 welded bends and a globe valve. Assuming an initial steam velocity of 30 m/s, determine the diameter of the main, the final steam pressure and the final steam velocity.

From steam tables the specific volume of steam at 0.17 MN/m² abs. pressure is found to be 1.03 m³/kg, then, from Eq (8.3),

$$d = 1.128 \sqrt{\frac{0.063 \times 1.03}{30}} = 0.053 \text{ m}$$

Nearest commercial diam. = 50 mm

or, alternatively, from Eq (8.4)

$$A^{-1} = \frac{30}{0.063 \times 1.03} = 460$$

and from Table 8.3 the nearest diameter = 50 mm. The actual value of $A^{-1} = 500$.

For a 50 mm-diam. pipe passing 0.063 kg/s we obtain the following data from Table 8.5:

$$\frac{Z_1 - Z_2}{l} = 0.007 \text{ and } l_e = 2.2$$

From Table 4.4 the velocity pressure factors are:

$$
\begin{array}{ll}
\text{Six 50 mm welded bends at } 0.3 = & 1.8 \\
\text{One 50 mm globe valve at } 5.0 \quad = & \underline{5.0} \\
& 6.8
\end{array}
$$

Then, total equivalent length of main (l)

$$= 50 + 6.8 \times 2.2 = 165 \text{ m}$$

and $Z_1 - Z_2 = 0.007 \times 165 = 1.155$

From Table 8.4 for $p_1 = 0.17$ MN/m² abs
                     $Z_1 = 2.78$
Therefore $Z_2 = 2.78 - 1.155 = 1.625$

and from Table 8.4 the final pressure

$$p_2 = 0.13 \text{ MN/m}^2 \text{ abs approx.}$$

From steam tables the specific volume of steam at 0.13 MN/m^2 abs pressure is found to be $1.033 \text{ m}^3/\text{kg}$ and, from Eq (8.4),

$$u = 500 \times 0.063 \times 1.033 = 32.5 \text{ m/s}$$

Ans.:   Diameter of main $= 50 \text{ mm}$
Final steam pressure $= 0.13 \text{ MN/m}^2$ abs
Final steam velocity $= 32.5 \text{ m/s}$

Example 8.3. A steam boiler supplies 12 unit heaters, each having an emission of 45 kW with steam at 0.28 MN/m^2 abs pressure and 0.9 dry. Assuming that the heat emission from the piping is additional to the unit heater emission and represents, as an average over the system, 10 per cent of the heat emission from the units, determine:

(*a*) The approximate steam flow rate per unit and for the system,
(*b*) the size of the pipe connection to each unit if the steam velocity is limited to 20 m/s

From steam tables

$$h_{fg} = 2\,171 \text{ kJ/kg}$$

and
$$v_g = 0.646 \text{ m}^3/\text{kg}$$

(*a*) Then, steam flow rate per unit

$$= \frac{45 \times 1.1}{0.9 \times 2\,171} = 0.0253 \text{ kg/s}$$

and for the system

$$12 \times 0.0253 = 0.3036 \text{ kg/s}$$

(*b*) From Eq (8.4)

$$A^{-1} = \frac{20}{0.0253 \times 0.646} = 1\,225$$

and from Table 8.3, $d = 32 \text{ mm}$

Example 8.4. The table below gives details of a heavy grade steam-piping layout that conveys saturated steam at an initial pressure of 0.48 MN/m^2 abs. A final pressure of about 0.45 MN/m^2 abs is required at the end of section 4.

(*a*) Determine the pipe size and mean velocity of each section and the actual final pressure of section 4.

(b) Determine the diameter of the branch pipe taken from the main at the end of section 1 if its length, including an allowance for local resistances, is 30 m. Also what will be the mean velocity of the steam in this branch.

Pipe section No.	1	2	3	4
Steam flow rate ($\dot{m}$), kg/s	0.38	0.13	0.11	0.025
Total equivalent length (l), m	50	30	15	30

(a) For

$$p_1 = 0.48 \text{ MN/m}^2 \text{ abs}; Z_1 = 20.61$$
$$p_2 = 0.45 \text{ MN/m}^2 \text{ abs}; Z_2 = 18.2$$
$$Z_1 - Z_2 = \overline{2.41}$$

Total equivalent length = 50 + 30 + 15 + 30 = 125 m

and
$$\frac{Z_1 - Z_2}{l} = \frac{2.41}{125} = 0.019$$

Using the data given in Tables 8.3, 8.4 and 8.5 and in the steam tables:

Pipe section No.	1	2	3	4
Steam flow rate ($\dot{m}$), kg/s	0.38	0.13	0.11	0.025
Initial design value of $\frac{Z_1 - Z_2}{l}$	0.019	0.019	0.019	0.019
Nearest diameter, mm	80	50	50	25
Actual value of $\frac{Z_1 - Z_2}{l}$	0.024	0.025	0.02	0.04
Equivalent length, m	50	30	15	30
Initial steam pressure (p_1), MN/m^2 abs	0.48	0.465	0.456	0.453
Z_1	20.61	19.41	18.66	18.36
$Z_1 - Z_2$	1.2	0.75	0.3	1.2
Z_2	19.41	18.66	18.36	17.16
Final steam pressure (p_2), MN/m^2 abs	0.465	0.456	0.453	0.434
Mean steam pressure (p_m), MN/m^2 abs approx.	0.47	0.46	0.45	0.44
Specific volume (v_g), m^3/kg	0.398	0.406	0.414	0.422
Velocity factor A^1	200	500	500	2 000
Mean steam velocity (u), m/s, Eq (8.4)	30	26	23	21

(b) For the branch pipe, assuming that the final pressure is the same as for section 4:

$$\text{Steam flow rate } (\dot{m}) = 0.38 - 0.13 = 0.25 \text{ kg/s}$$
$$\text{Initial pressure } (p_1) = 0.465 \text{ MN/m}^2 \text{ abs}$$
$$\text{Final pressure } (p_2) = 0.434 \text{ MN/m}^2 \text{ abs}$$
$$Z_1 = 19.41$$
$$Z_2 = 17.16$$
$$Z_1 - Z_2 = 2.25$$
$$\text{Total equivalent length} = 30 \text{ m}$$

$$\frac{Z_1 - Z_2}{l} = \frac{2.25}{30} = 0.075$$

Nearest diameter = 50 mm

Actual value of $\dfrac{Z_1 - Z_2}{l} = 0.094$

then $\qquad\qquad Z_1 - Z_2 = 30 \times 0.094 = 2.82$

and $\qquad\qquad Z_2 = 19.41 - 2.82 = 16.59$

Actual final pressure	$(p_2) = 0.43$ MN/m² abs approx.
Mean pressure	$(p_m) = 0.45$ MN/m² abs approx.
Specific volume	$(v_g) = 0.414$ m³/kg
Velocity factor	$(A^{-1}) = 500$
Mean steam velocity	$(u) = 52$ m/s

Since this velocity is too high for saturated steam, it will be necessary to increase the diameter of the branch pipe to 65 mm.
Then:

Actual value of $\dfrac{Z_1 - Z_2}{l} = 0.025$

$$Z_1 - Z_2 = 30 \times 0.025 = 0.75$$

$$Z_2 = 19.41 - 0.75 = 18.66$$

and actual final pressure $\qquad p_2 = 0.455$ MN/m² abs

which is in excess of the required final pressure by $0.455 - 0.434 = 0.021$ MN/m²; this may be absorbed by valve adjustment at the unit, by steam-pressure reducing valve or by orifice plate. For the 65 mm pipe the mean pressure will be about 0.46 MN/m² abs, having a specific volume of 0.406 m³/kg; the velocity factor from Table 8.3 is 300. The mean steam velocity will then be from Eq (8.4) 30 m/s which is satisfactory.

Summary of results:

Pipe section No.	1	2	3	4	Branch
Diameter, mm	80	50	50	25	65
Mean steam velocity, m/s	30	26	23	21	30

Examples 8.5. The plan layout of a saturated steam-distribution system is shown diagrammatically in Fig. 8.2. Using the data given below size all the pipes and determine the steam pressure at points *a*, *b* and *c*.

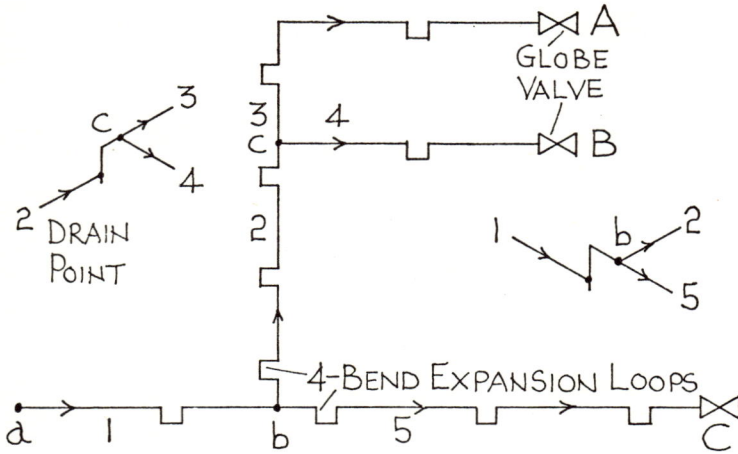

Fig. 8.2. Steam-pipe sizing example.

Data:

Tubing: Heavy grade with welded fittings.

Unit	A	B	C
Steam pressure required, MN/m² abs (p) .	0.4	0.4	0.5
Steam flow rate, kg/s $(\dot{m})$	0.08	0.05	0.06

Pipe section No. ..	1	2	3	4	5
Length, m	70	170	100	70	130

In this case the pipe sizing may be carried out by either: (*a*) fixing the steam velocity, or (*b*) using an average pressure drop of, say, $200 \frac{\text{N}}{\text{m}^2}$ m of equivalent length. Since the steam pressure at *a*, *b* and *c* is required, it will be more straightforward to size pipe sections 1, 2 and 3 on a fixed velocity and then to work back from diameter to pressure drop, and hence initial pressure of each pipe.

Consider pipe No. 3 serving unit A:
From steam tables for $p = 0.4$ MN/m² abs

$$v_g = 0.46 \text{ m}^3/\text{kg}$$

Then from Eq (8.4), assuming a final steam velocity of 30 m/s

$$A^{-1} = \frac{30}{0.08 \times 0.46} = 815$$

and from Table 8.3 the nearest diameter is found to be 40 mm. From Table 8.5 for a steam flow rate of 0.08 kg/s

$$\frac{Z_1 - Z_2}{l} = 0.036 \text{ and } l_e = 1.7$$

From Table 4.4 the velocity pressure factors are:

$$
\begin{aligned}
\text{Nine 40 mm welded bends at } 0.3 &= 2.7 \\
\text{One 40 mm globe valve at } 5.0 \quad &= \underline{5.0} \\
& \underline{7.7}
\end{aligned}
$$

Then the total equivalent length will be

$$100 + (7.7 \times 1.7) = 113 \text{ m}$$

and $\qquad Z_1 - Z_2 = 113 \times 0.036 = 4.07$

From Table 8.4 for $\qquad p_2 = 0.4 \text{ MN/m}^2 \text{ abs}$

$$Z_2 = 14.5$$

Then $\qquad Z_1 = 14.5 + 4.07 = 18.57$

and $\qquad p_1 = 0.455 \text{ MN/m}^2 \text{ abs}$

That is, pressure at "c" = 0.455 MN/m² abs.

Consider pipe No. 2 having a final pressure of 0.455 MN/m² abs. From steam tables

$$v_g = 0.41 \text{ m}^3/\text{kg}$$

Then from Eq (8.4), assuming a final steam velocity of 30 m/s,

$$A^{-1} = \frac{30}{(0.08 + 0.05)0.41} = 563$$

and from Table 8.3 the nearest pipe diameter is found to be 50 mm. From Table 8.5 for a steam flow rate of 0.13 kg/s

$$\frac{Z_1 - Z_2}{l} = 0.028 \text{ and } l_e = 2.3$$

The velocity pressure factors are:

$$
\begin{aligned}
\text{Thirteen 50 mm welded bends at } 0.3 &= 3.9 \\
\text{One 50 mm tee, 90}^\circ, \text{ at} \quad 0.5 + 0.4 &= 0.9 \\
\text{One 50 mm tee, run, at} \quad 0.2 + 0.4 &= \underline{0.6} \\
& \underline{5.4}
\end{aligned}
$$

Then total equivalent length will be:

$$170 + (5.4 \times 2.3) = 182 \text{ m}$$

and $\qquad Z_1 - Z_2 = 182 \times 0.028 = 5.09$

since $\qquad Z_2 = 18.57$

$\qquad Z_1 = 18.57 + 5.09 = 23.66$

and $\qquad p_1 = 0.515 \text{ MN/m}^2 \text{ abs}$

That is, pressure at "*b*" = 0.515 MN/m² abs.
Consider pipe No. 1 having a final pressure of 0.515 MN/m² abs and proceeding as above for pipes No. 2 and 3.
From steam tables for $p = 0.515$ MN/m² abs.

$$v_g = 0.37 \text{ m}^3/\text{kg}$$

$$A^{-1} = \frac{30}{(0.08 + 0.05 + 0.06)0.37} = 427$$

Nearest diameter = 50 mm, for which $A^{-1} = 500$

and the actual velocity will be, from Eq (8.4),

$$u = 0.19 \times 0.37 \times 500 = 35 \text{ m/s}$$

which is acceptable for a horizontal main.
From Table 8.5 for a steam flow rate of 0.19 kg/s

$$\frac{Z_1 - Z_2}{l} = 0.056 \text{ and } l_e = 2.4$$

The velocity pressure factors are, including the drain point at "*b*",

$$\begin{aligned}\text{Five 50 mm welded bends at } 0.3 &= 1.5 \\ \text{Two 50 mm } 90° \text{ tees at } 0.5 + 0.4 &= 1.8 \\ &= \overline{3.3}\end{aligned}$$

Then, total equivalent length will be

$$70 + (3.3 \times 2.4) = 78 \text{ m}$$

and $\qquad Z_1 - Z_2 = 78 \times 0.056 = 4.37$

since $\qquad Z_2 = 23.66$

$\qquad Z_1 = 23.66 + 4.37 = 28.03$

and $\qquad p_1 = 0.56 \text{ MN/m}^2 \text{ abs}$

That is, pressure at "*a*" = 0.56 MN/m² abs.
Consider pipe No. 4, for which

$$\begin{aligned}p_1 &= 0.455 \text{ MN/m}^2 \text{ abs and } Z_1 = 18.57 \\ p_2 &= 0.4 \text{ MN/m}^2 \text{ abs and } Z_2 = 14.5 \\ Z_1 - Z_2 &= \overline{4.07}\end{aligned}$$

Allowing, say, 15 per cent extra length for local resistances

$$\frac{Z_1 - Z_2}{l} = \frac{4.07}{1.15 \times 70} = 0.05$$

Then, from Table 8.5 for a steam flow rate of 0.05 kg/s, the nearest pipe diameter is found to be 32 mm, for which the actual

$$\frac{Z_1 - Z_2}{l} = 0.035 \text{ and } l_e = 1.4$$

The velocity pressure factors are:

Four 32 mm welded bends at 0.3 = 1.2
One 32 mm globe valve at 5.0   = 5.0
                                = 6.2

Then, total equivalent length will be

$$70 + (6.2 \times 1.4) = 79 \text{ m}$$

and $Z_1 - Z_2 = 79 \times 0.035 = 2.77$

since $Z_1 = 18.57$

$$Z_2 = 18.57 - 2.77 = 15.8$$

and $p_2 = 0.418 \text{ MN/m}^2 \text{ abs}$

This pressure is greater than the required final pressure, which may be obtained by including a resistance equivalent to $0.418 - 0.4 = 0.018 \text{ MN/m}^2$ in the pipe or by valve throttling.

For 32 mm pipe $A^{-1} = 1\,200$

and since $v_g = 0.41 \text{ m}^3/\text{kg}$, the initial steam velocity will be

$$u = 0.05 \times 0.41 \times 1\,200 = 25 \text{ m/s}$$

Consider pipe No. 5, for which

$p_1 = 0.515 \text{ MN/m}^2 \text{ abs and } Z_1 = 23.66$
$p_2 = 0.5 \text{ MN/m}^2 \text{ abs and } Z_2 \quad = 22.30$
$\qquad\qquad Z_1 - Z_2 \qquad\qquad = \overline{1.36}$

Allowing 15 per cent extra length for local resistances

$$\frac{Z_1 - Z_2}{l} = \frac{1.36}{1.15 \times 130} = 0.009$$

Then, from Table 8.5, for a steam flow rate of 0.06 kg/s the nearest pipe diameter is found to be 50 mm, for which the actual

$$\frac{Z_1 - Z_2}{l} = 0.0063 \text{ and } l_e = 2.2$$

The velocity pressure factors are:

Twelve 50 mm welded bends at 0.3 = 3.6
One 50 mm globe valve at 5.0   = 5.0
                                 = 8.6

Then, total equivalent length will be

$$130 + (8.6 \times 2.2) = 149 \text{ m}$$

and $\qquad Z_1 - Z_2 = 149 \times 0.0063 = 0.94$

since $\qquad Z_1 = 23.66$

$$Z_2 = 23.66 - 0.94 = 22.72$$

and $\qquad p_2 = 0.505 \text{ MN/m}^2 \text{ abs}$

which is only slightly above the required final pressure.

For 50 mm pipe, $\qquad A^{-1} = 500$
and since $v_g = 0.37 \text{ m}^3/\text{kg}$, the initial steam velocity will be

$$u = 0.06 \times 0.37 \times 500 = 11 \text{ m/s}$$

Summary of results:

Pipe No.	1	2	3	4	5
Diameter, mm	50	50	40	32	50

Point			a	b	c
Steam pressure, MN/m² abs			0.56	0.515	0.455

Example 8.6. A 50 mm-diameter heavy grade steam main conveys 0.05 kg/s of saturated steam at a mean pressure of 0.3 MN/m² abs. The length of the main, including an allowance for local resistances, is 150 m.

(*a*) Determine the initial and final steam pressures and mean velocity.

(*b*) It is required to increase the steam flow rate to 0.11 kg/s. Using the final pressure found in part (*a*), determine the new initial pressure and mean velocity.

(*a*) $p_m = 0.3 \text{ MN/m}^2$ abs and $Z_m = 8.32$
From Table 8.5 for a 50 mm pipe passing 0.05 kg/s

$$\frac{Z_1 - Z_2}{l} = 0.0045$$

Then, since $\qquad l = 150 \text{ m}$

$$Z_1 - Z_2 = 150 \times 0.0045 = 0.68$$

$$Z_1 = 8.32 + \frac{0.68}{2} = 8.66$$

and $\qquad Z_2 = 8.32 - \frac{0.68}{2} = 7.98$

From which

$$\text{Initial pressure, } p_1 = 0.305 \text{ MN/m}^2 \text{ abs}$$
$$\text{Final pressure, } p_2 = 0.295 \text{ '' '' ''}$$

From steam tables for the initial pressure

$$v_g = 0.6 \text{ m}^3/\text{kg}$$

For a 50 mm pipe $A^{-1} = 500$

Then $u_1 = 0.05 \times 0.6 \times 500 = 15 \text{ m/s}$

(*b*) From Table 8.5 for a 50 mm pipe passing 0.11 kg/s

$$\frac{Z_1 - Z_2}{l} = 0.02$$

Then $Z_1 - Z_2 = 0.02 \times 150 = 3.0$

From part (*a*) $Z_2 = 7.98$

∴ $Z_1 = 7.98 + 3.0 = 10.98$

from which $p_1 = 0.345 \text{ MN/m}^2 \text{ abs}$

Mean pressure $= \frac{1}{2}(0.345 + 0.295) = 0.32 \text{ MN/m}^2 \text{ abs}$

for which $v_g = 0.56 \text{ m}^3/\text{kg}$

Then, since $A^{-1} = 500$

$$u = 0.11 \times 0.56 \times 500 = 31 \text{ m/s}$$

Problems

1. A steam-heated air heater battery having a heat output of 103 kW is supplied with steam at 0.207 MN/m² abs pressure and 0.9 dry. The steam main is horizontal and 61 m long and contains eight welded bends, one drain point and one globe valve. If the battery uses only the latent heat in the steam and the final velocity is to be about 36 m/s, neglect the effect of heat emission from the main and determine:

(*a*) the diameter of the main;
(*b*) the initial steam pressure;
(*c*) the initial steam velocity.

Ans.: (*a*) 40 mm; (*b*) 0.246 MN/m² abs; (*c*) 31.5 m/s

2. Determine the diameter of the following horizontal steam main:

Length   = 91 m
Local resistances   = 8 welded bends and 1 globe valve
Steam flow rate   = 0.038 kg/s
Initial pressure (p_1) = 0.379 MN/m² abs
Final pressure (p_2) = 0.345   ”   ”   ”

Ans.: d = 32 mm giving p_2 = 0.33 MN/m² abs, u_1 = 20 m/s and u_2 = 22 m/s
u_2 = 22 m/s.

3. An existing 125 mm-diameter high-pressure steam main carries 1.0 kg/s
of dry saturated steam having a mean pressure of 0.516 MN/m² abs over a
distance of 152 m.

(a)   (i) What are the pressures at the entry and exit to the main?
       (ii) State the carrying capacity of the main if the actual pressure
            drop is doubled.
      (iii) Find the mean steam velocities in both cases.

The fittings and valves in the main are given as:

tees, 4; expansion loops, 3; valves, 2.

(b) Make a sketch of a typical drain point in such a main.   (IHVE)

Ans.: (i) 0.565 and 0.496 MN/m² abs respectively.
       (ii) 1.39 kg/s approx.
      (iii) 28 and 38 m/s respectively.

4. The table below gives details of a heavy grade steam-piping layout that
conveys saturated steam at an initial pressure of 0.31 MN/m² abs.

(a) Assuming a steam velocity of 30 m/s throughout, determine the
approximate pipe size of each section.

(b) The branch pipe at the end of section 1 has an equivalent length of
15.25 m and a terminal pressure of 0.303 MN/m² abs. Assuming that
section 1 is 10 m long and contains a boiler exit, four welded bends and a
globe valve, determine the diameter of the branch pipe.

Pipe reaction No.	1	2	3	
Steam flow rate, kg/s	0.254	0.108	0.066	(NC)

Ans.: 80 mm, 50 mm, 40 mm and 65 mm respectively.

Pipe No.	1	2	3	4	5
Diameter, mm.	50	20	50	32	32

Point		a	b	c
Steam pressure, MN/m² abs		0.386	0.375	0.352

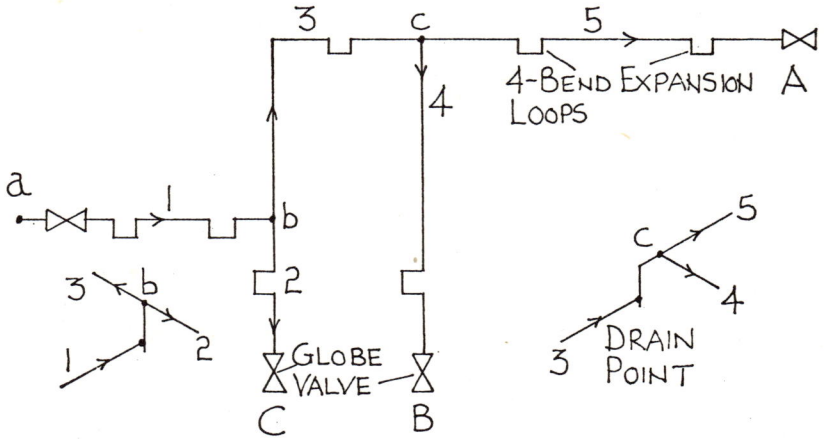

Fig. 8.3. A steam distribution system.

5. The steam distribution system shown diagrammatically in Fig. 8.3 is made up from Heavy Grade tube and welded fittings. Using the data tabulated below, size all the pipes and determine the steam pressure at points *a*, *b* and *c*.   (IHVE)

Data:

Terminal	A	B	C
Steam pressure required, MN/m^2 abs	0.303	0.241	0.303
Steam flow rate, kg/s	0.05	0.026	0.017

Pipe No.	1	2	3	4	5
Length, m	61	30.5	61	61	61

Ans.:

Pipe No.	1	2	3	4	5
Diameter, mm	50	20	50	32	32

Point	*a*	*b*	*c*
p, MN/m^2 abs	0.28	0.26	0.25

9: Hot-Water Supply

Introduction

Hot-water supply systems for domestic and industrial purposes are required to supply water at the specified temperature and in sufficient quantities to meet the demand. The choice of a system depends upon the following main factors:

(i) Number and type of draw-off points to be served and whether they are grouped together or widely distributed.
(ii) Whether the demand is continuous or intermittent.
(iii) Whether the system is to be combined with space heating.
(iv) Nature of the water supply.

The materials used also depend upon the nature of the water, which may be corrosive or have scale or sludge-forming tendencies.

Electrolytic corrosion may occur whenever dissimilar metals are in contact in the presence of water therefore if copper pipes are used in conjunction with galvanized steel tanks and storage cylinders or vice versa, they should be separated by a non-metallic coupling. If the water happens to be cupro-solvent however, copper from the pipes and copper alloy fittings will be deposited on any galvanized steel surfaces leading to severe pitting. If the water is hard a protective layer of scale may be rapidly deposited and may protect the underlying metal against electrolytic corrosion. Excessive scale formation from hard water may be minimized by adding a small quantity of a metaphosphate complex to the cold water supply to the system. Soft waters are not normally scale forming but may cause chemical corrosion of unprotected iron and steel and in some cases may also be plumbo-solvent. Non-ferrous equipment should be used wherever possible and cast-iron boilers should be protected by, say, vitreous enamel to prevent discoloration of the hot water. Some water has a certain relationship between its chloride content and temporary hardness which may lead to a form of corrosion known as dezincification in which the brass of duplex brass hot pressed fittings is converted to porous copper resulting in water leakage and ultimate failure of the fittings. Before selecting materials to be used in any installation information concerning the properties of the local water supply should be obtained from the relevant water supply authority. The amount of scale or sludge in the system depends not only on the quality of the water but also on the maximum temperature attained, and for most waters it is recommended that the temperature shall not exceed 65°C. For this reason only indirect systems, in which the water in the primary circuit is unchanged and can if necessary be at a temperature in excess of 65°C, should be used.

Typical water temperatures, capacities and discharge rates are given in Table 9.1.

Consumption, Storage and Boiler Power

The principle of incorporating storage vessels in hot water systems serving a number of draw-off points is based on the fact that the stored water can be used to meet periods of peak demand that are significantly greater than the recovery capacity of the water heater. In this way the hourly energy consumption will be minimized and may be spread throughout the day; this is particularly convenient when off-peak electricity is used. Since the amount of storage to be provided to meet a given daily demand will depend upon the rate of supply of heat it follows that there exists an infinite variety of these two variables. The most economical combination should be used and this may be determined only from a detailed cost analysis; instantaneous water heaters may be suitable in some cases.

In the absence of specific information or instructions regarding conditions of maximum consumption and draw-off the data given in Table 9.2 may be adopted as the basis of design. This data is based on B.S. CP: 342, which should be consulted for further information.

Simultaneous Demand

The simultaneous demand from a number of draw-off points from which water will be discharged intermittently may be found with sufficient accuracy from the expression:*

$$m \simeq n\left(\frac{\cdot t}{T}\right) + 1.8 \sqrt{2n\left(\frac{t}{T}\right)\left(1 - \frac{t}{T}\right)} \qquad (9.1)$$

where m = number of draw-off points simultaneously discharging, taken to the nearest whole number

n = total number of draw-off points

t = average time draw-off point is discharging for each occasion of use

T = average time between occasions of use.

* L. C. BULL, "Simultaneous Demand from a Number of Draw-off Points" *J.I.H.V.E.*, **23**, pp. 445–451 (1956).

TABLE 9.1

Type of draw-off	Temperature, °C	Capacity, litre	Discharge rate, l/s
Basin	45−60	4.5−9	0.1
Bath	45−60	110−140	0.4−0.6
Shower rose	45		0.3−0.6
Sink	60	20−45	0.3

TABLE 9.2

Hot Water Demand, Storage and Boiler Power

Hot water demand given in litre per person per day of 24 hours *on day of heaviest demand* during week. The boiler must be rated at not more than 19 kW/m² of boiler heating surface and should provide for the requirements set out below plus any heat losses from towel rails, coils and circulating pipes.

Building	Maximum daily demand litre per person	Storage litre per person	Boiler power kW per person
Colleges and schools:			
Boarding	115	23	0.75
Day	15	4.5	0.1
Dwelling-houses:			
Medium rental	115	46	0.9
High rental	140	46	1.2
Factories	15	4.5	0.1
Flats, blocks of:			
Low rental	70	23	0.45
Medium rental	115	32	0.75
High rental	140	32	0.9
Hospitals:			
General	140	27	1.5
Infectious	230	46	1.5
Maternity	230	32	2.0
Nurses' homes	140	46	0.9
Hotels:			
First class	140	46	1.2
Average	115	36	0.9
Offices...................	15	4.5	0.1

Note.

1. Low Rental = houses and flats not exceeding 90 m² in area.
   Medium Rental = houses and flats exceeding 90 m² in area but less than 140 m².
   High Rental = houses and flats exceeding 140 m² in area.
2. For a dwelling of about 90 m² floor area there should be:

   (*a*) A storage vessel of not less than 160 l actual capacity to enable 15 l of hot water to be drawn off at the kitchen sink at a temperature of not less than 60°C, immediately prior to two baths being taken in succession.

   (*b*) Means for warming the linen cupboard. Generally, sufficient heat is given off by an insulated hot tank to maintain a cupboard warm and dry.

   (*c*) A heated towel-airing pipe or heated towel rail.

(*d*) A water heater capable of heating the effective contents of the storage vessel from 10°C to 65°C in 4 h, while heating at the same time the other connected loads such as a towel rail.

For appliances having a probability of 0.2, Eq (9.1) becomes:

$$m = 0.2n + \sqrt{n} \qquad (9.2)$$

The solution of Eq (9.2), with *m* taken to the nearest whole number and expressed as a percentage of *n*, is given in Table 9.3.

TABLE 9.3

n	*m*	$100\,\dfrac{m}{n}$	*n*	*m*	$100\,\dfrac{m}{n}$
2	2	100	10	5	50
3	2	67	15	7	47
4	3	75	20	8	40
5	3	60	30	11	37
6	4	67	40	14	35
7	4	57	50	17	34
8	4	50	60	20	33
9	5	56	70	22	32

n = total number of draw-off points.
m = number of points simultaneously discharging.

For systems containing a mixture of points having different probability ratios, the probable simultaneous demand of the combined system may be found by determining the total equivalent number of points to a common probability.

The value of *t*/T varies considerably for systems containing a mixture of baths, basins and sinks depending not only on the relative number of each type of appliance but also on the frequency of usage. In such cases it is convenient to use the concept of 'demand units' which take into account both the flow rate at the appliance and its probable usage. Comparative demand units for standard appliances are given for different types of buildings in Table 9.4 and converted to water flow rate in Fig. 9.1.

TABLE 9.4

Comparative Demand Units

Application	Basin		Bath		Sink	
	Hot	Cold	Hot	Cold	Hot	Cold
Hotels, industrial buildings, offices, schools and similar. (*t*/T 0.2 – 0.3)	1	0.7	5	3	6	2.5
Flats, restaurants, theatres and similar. (*t*/T 0.1 – 0.2)	0.5	0.4	2	1.2	3	1.3
Where there are more than 50 draw-off points. (*t*/T 0.05 – 0.1)	0.3	0.2	1	0.6	2	0.7

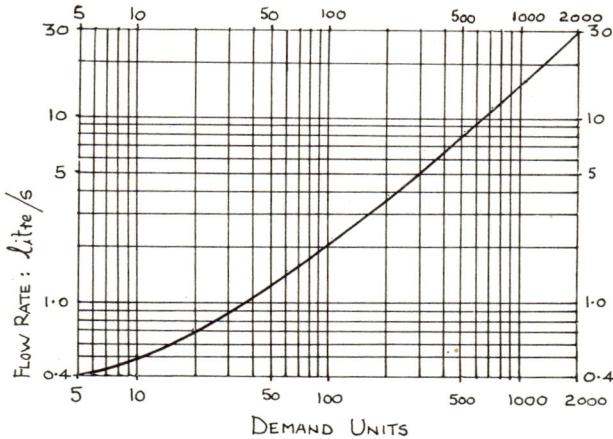

Fig. 9.1.

Pipe-sizing. The secondary circulation consists of:

(*a*) a flow pipe sized to give the required rate of outflow at the draw-off points.

(*b*) a return pipe capable of carrying the water flow rate determined from the total heat emission from the system and the temperature drop across the circuit.

The return pipe is included to enable a circulation to take place at times when all the draw-off points are closed, thus ensuring that hot water is always near the points. When draw-off takes place circulation ceases and water for outflow may flow from the storage vessel along both the flow and return pipes. This is particularly true for natural circulation ring-main systems, but the effect is generally ignored.

The diameters selected for the flow piping should therefore be capable of passing the maximum simultaneous demand, and will depend upon the available pressure and the circuit length. For simplicity and to avoid additional velocity pressure calculations, it is usual to calculate the available pressure, by considering the vertical distance from the bottom of the feed tank to the level of the draw-off only, leaving the pressure due to the water in the tank to ensure discharge at the draw-off point. The pressure drop across the draw-off appliance is usually known and should be deducted from the available pressure leaving the balance to overcome the resistance offered to flow by the pipes and remaining local resistances. The index circuit, i.e. the circuit which has the lowest value of pressure available divided by the equivalent length from the feed tank to the index draw-off, should be sized first. The remaining branch pipes may then be sized and balanced generally in accordance with the pipe sizing procedures previously laid down. It should be noted that while the cold feed to the system is part of the index circuit it should be sized using tables for water flowing at 10°C. Pipe sizing data for hot and cold water flowing in copper, galvanized and cast-iron pipes are given in Part C of the current edition of the Institution of Heating and Ventilating Engineers Guide to Practice.

Since the return pipe is part of a hot-water circulation, it should be sized generally in accordance with the procedures previously laid down for hot-water heating systems. This applies also to the primary circuit between the boiler and the storage cylinder.

Example 9.1. Calculate the probable simultaneous demand from the hot-water supply system detailed below:

Number of draw-off points	(n)	30
Discharge rate per point		0.75 l/s
Time draw-off point is discharging	(t)	3 min
Average time between occasions of use	(T)	30 min

Using Eq (9.1),

$$m \simeq 30 \times (\tfrac{3}{30}) + 1.8 \sqrt{2 \times 30 \times (\tfrac{3}{30})(1 - \tfrac{3}{30})}$$

from which, to the nearest whole number

$$m = 7$$

The probable simultaneous demand will therefore be:

$$7 \times 0.75 = 5.25 \text{ l/s}$$

Example 9.2. Calculate the probable maximum rate of outflow from a storage vessel that serves the following hot-water supply system:

Type of point	A	B
Number of points (n)	30	20
Discharge rate per point, l/s	0.2	0.8
Probability $\left(\dfrac{t}{T}\right)$	0.2	0.3

Consider the twenty "B" type points, then from Eq (9.1),

$$m \simeq 20 \times 0.3 + 1.8 \sqrt{2 \times 20 \times 0.3(1 - 0.3)}$$

from which, to the nearest whole number,

$$m = 11$$

and the rate of discharge will be

$$11 \times 0.8 = 8.8 \text{ l/s}$$

which is equivalent to $\dfrac{8.8}{0.2} = 44$ "A" type points discharging.

Then, since $\frac{t}{T}$ for the A type points = 0.2, using the simplified Eq (9.2),

$$44 = 0.2n + \sqrt{n}$$

Let $\qquad x^2 = n$ and $x = \sqrt{n}$

Then $\qquad 44 = 0.2x^2 + x$

Rearranging, $\qquad 0.2x^2 + x - 44 = 0$

and solving the quadratic by formula,

$$x = \frac{-1 \pm \sqrt{1 + 4 \times 0.2 \times 44}}{2 \times 0.2}$$

from which $\qquad x = 12.5$

therefore $\qquad n = 12.5^2 = 156$

That is, 156 A type points are equivalent to 20 B type. The complete system may therefore be taken as equivalent to 156 + 30 = 186 A type points and then from Eq (9.2)

$$m = 0.2 \times 186 + \sqrt{186}$$
$$= 51$$

The probable maximum rate of outflow from the storage vessel will then be:

$$51 \times 0.2 = 10.2 \text{ l/s}$$

Example 9.3. The secondary outflow piping of a hot water supply system for a block of 12 medium rental flats is shown diagrammatically in Fig. 9.2.
(*a*) Determine the water flow rate for each pipe of the index circuit.
(*b*) Determine the capacity of the storage vessel and the net boiler power.

(*a*) Using the data given in Table 9.4, the total number of demand units per flat will be:

$$0.5(\text{basin}) + 2(\text{bath}) + 3(\text{sink}) = 5.5$$

and from Fig. 9.1 the water flow rate is found to be 0.42 l/s. The total number of demand units and the corresponding water flow rates for each pipe will be:—

Pipe number	1	2	3	4	5	6
Demand units	66	66	44	22	11	5.5
Flow rate, l/s	1.6	1.6	1.2	0.76	0.52	0.42

(*b*) From Table 9.2, for medium rental flats:

Storage, litre per person = 32
Boiler power, kW per person = 0.75

FEED TANK

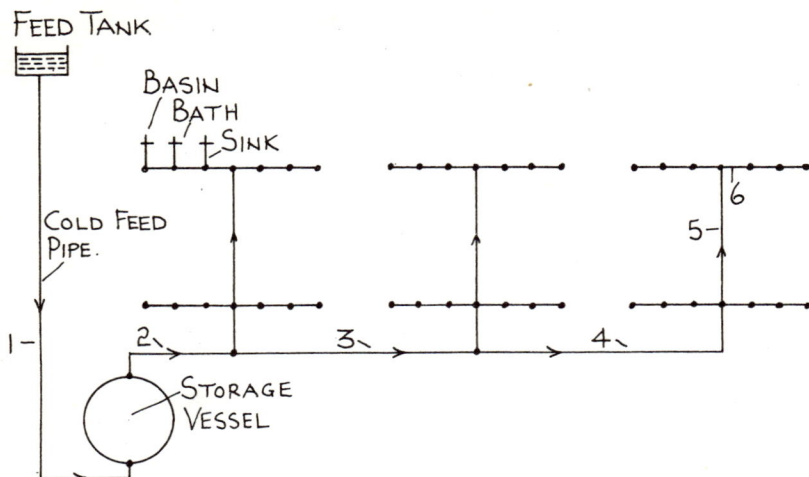

Fig. 9.2. Secondary outflow piping.

Then, assuming 3 persons per flat,

Storage to be allowed = 3 × 12 × 32 = 1 152 litre

and

Boiler power,   = 3 × 12 × 0.75
                     = 27 kW

This is the net boiler power, and excludes circulation losses and margins.

Example 9.4. A steam-heated indirect hot-water storage cylinder contains 2 300 kg of water initially at 10°C. Using the data listed below, determine the temperature of the water after a 4-hour heating-up period. Assume negligible heat losses and thorough mixing of the stored water.

Data:

Heating surface area   = 0.37 m^2
Thermal transmittance of coil = 0.57 kW/m^2°C
Temperature of the steam   = 120°C
Specific heat capacity of water = 4.2 kJ/kg°C

Let t = temperature of water at time θ

and dt = rise in temperature in interval $d\theta$

The, since the heat transferred by the coil is equal to the heat gained by the water

$$0.57 \times 0.37(120 - t)d\theta = 2\,300 \times 4.2\,dt$$

integrating by separation of variables

$$\int_0^\theta d\theta = \frac{2\,300 \times 4.2}{0.57 \times 0.37} \int_{10}^t \frac{dt}{120 - t}$$

$$\theta = 46\,000\,[\log_e (120 - t)]_{10}^t$$

$$= 46\,000\,\log_e \frac{110}{120 - t}$$

for

$$\theta = 4 \text{ hours} = 14\,400 \text{ s}$$

$$14\,400 = 46\,000\,\log_e \frac{110}{120 - t}$$

or

$$0.313 = \log_e \frac{110}{120 - t}$$

and

$$\frac{110}{120 - t} = e^{0.313}$$

$$= 1.37$$

hence

$$t = 120 - \frac{110}{1.37}$$

$$= 40°C$$

Example 9.5. It is required to raise the temperature of 2 300 kg of water from 10°C to 65°C in 3 h in the steam-heated storage vessel detailed below. Assuming negligible heat losses and thorough mixing of the stored water, determine the steam temperature

Heating surface area   1.0 m²
Thermal transmittance of coil   680 W/m²°C

From the previous example it is seen that

$$\theta = \frac{mc}{UA} \log_e \frac{t_s - t_1}{t_s - t_2}$$

where θ = time
c = specific heat capacity
m = mass of water
U = thermal transmittance of heating surface
A = heating surface area
t_s = temperature of steam
t_1 = initial temperature of water
t_2 = final temperature of water.

Then $$3 \times 3\,600 = \frac{2\,300 \times 4.2 \times 10^3}{680} \log_e \frac{t_s - 10}{t_s - 65}$$

from which $$\log_e \frac{t_s - 10}{t_s - 65} = 0.76$$

and $$\frac{t_s - 10}{t_s - 65} = 2.139$$

then $$t_s = 113°C$$

Problems

1. Calculate the probable simultaneous demand from the following hot-water service system:

Number of points	50
Discharge rate per point	0.76 l/s
Time draw-off point is discharging (t)	4
Average time between occasions of use (T)	40

Ans.: 7.88 l/s.

2. A 2 720 kg indirect hot-water storage cylinder is filled with water at 10°C. The coil has 0.46 m² of heating surface and is supplied with steam at 116°C. Assuming a constant thermal transmittance for the coil of 680 W/m²°C, plot a curve to show the rise of temperature of the water during a 6-h heating-up period. Clearly state any assumptions which may be necessary.

Ans.:

t	27	38	49	60
θ	1.7	3.07	4.57	6.41

3. A pipe in a hot-water service system serves 100 draw-off points. On each occasion of use 40 of these points discharge at 0.38 l/s for an average time of 5 min the average time between occasions of use being 20 min. The remainder discharge at 0.11 l/s for an average time of 3 min, the average time between occasions of use being 10 min. Determine the flow rate on which the pipe should be sized.

Ans.: 8.87 l/s.

4. Calculate the probable maximum rate of out-flow from the following hot-water supply system:

	A	B
Type of point	A	B
Number of points	50	20
Discharge rate per point	0.11	0.38
Time draw-off point is discharging (t)	3	5
Average time between occasions of use (T)	10	25

Ans.: 4.93 l/s